D0938868

Saving Energy in
Manufacturing

Saving Energy in Manufacturing

The Post-Embargo Record

John G. Myers
Southern Illinois University

Leonard Nakamura
The Conference Board

A project jointly sponsored by The Conference Board,
the Alliance to Save Energy and the Ford Foundation

Ballinger Publishing Company • Cambridge, Massachusetts
A Subsidiary of J.B. Lippincott Company

This book is printed on recycled paper.

International Standard Book Number: 0−88410−082−0

Library of Congress Catalog Card Number: 78−7586

Printed in the United States of America

Library of Congress Cataloging in Publication Data

Myers, John G
  Saving energy in manufacturing.

  Includes bibliographical references.
  1. Factories—Energy conservation. I. Nakamura, Leonard I., joint author. II. Title.
TJ163.5.F3M93   333.7   78−7586
ISBN 0−88410−082−0

Contents

List of Figures

List of Tables

Foreword

Nearly five years have elapsed since the Arab members of the Organization of Petroleum Exporting Countries declared an embargo, in October 1973, on shipments of petroleum to the United States as well as to the Netherlands, Portugal, and South Africa. That action, together with the subsequent rapid and extensive increases in world petroleum prices, created the "energy crisis." This book examines the developments in U.S. manufacturing industries since the embargo.

An obvious response of manufacturers to higher energy prices and to energy supply interruptions would be to seek to reduce energy use per unit of product. In the years since 1973, however, the nation has experienced severe inflation, a serious recession, and only partial recovery. These developments have not aided the ability of manufacturing firms to make the often quite expensive adjustments necessary to save energy.

We described developments in energy use from 1947 to 1967 in a study published in 1974.[a] In this book we examine subsequent events, concentrating on the 1974–1976 period, in search of answers to the questions: Has energy saving accelerated since the embargo? In what industries were recent savings concentrated? How were the savings achieved? What were the principal impediments to saving energy during the post-embargo period? What were the main aids to saving energy? The answer to the first question is positive—there was an

[a]J.G. Myers et al., *Energy Consumption in Manufacturing* (Cambridge, Mass.: Ballinger Publishing Company, 1974).

acceleration in energy saving. Answers to the other questions are summarized in Chapter 1 and are presented in greater detail in the rest of the text.

Our investigation was encouraged by Walter A. Hamilton of The Conference Board, Richard L. McGraw of the Alliance to Save Energy, and Allan G. Pulsipher of the Ford Foundation. Their interest in the topic and financial support are gratefully acknowledged.

The cornerstone of any study of this type is timely data. We benefited from the excellent cooperation provided by employees of the Bureau of the Census, the Bureau of Mines, and the Department of Energy. In particular we wish to thank Ruth Runyan, Edward A. Robinson, Dennis Wagner, and John McNamee of the Bureau of the Census; Norman Wingard of the Bureau of Mines; and Jim Alexander and Harry Gresham of the Department of Energy (Oak Ridge) for their help in providing data and patiently answering our many questions.

We also received extensive assistance in our work from business. Of the many persons who contributed valuable data and technical advice, we are especially indebted to A. Bruce Burns and Ronald S. Wishart of Union Carbide, Jeffrey Duke of the American Paper Institute, Stewart G. Fletcher of the American Iron and Steel Institute, Thomas R. O'Connor of the Portland Cement Association. and Allen Sheldon and J. Lee O'Nan of ALCOA.

In the course of the study we received skillful research assistance from Suk-mo Koo, Mark Claycomb, Linda Modica, and Amy Halperin. The text was made more readable and accurate by the editorial assistance of Bernard A. Gelb, Amy Halperin, and Linda Modica. One of our greatest debts is to Lora Peralto, typist extraordinary.

Helpful comments on an early draft of this book were provided by a number of persons and organizations. Particularly valuable were those received from Joel Darmstadter of Resources for the Future, Robert A. Leone of Harvard University, Denis M. Slavich of Betchel Corporation, David R. Dilley and Marvin L. Hughes of U.S. Steel, J.P. Hammond of Standard Oil (Indiana), Lawrence M. Woods of Mobil Oil Corporation, Sara Lee Soder of SRI International, Brian Sullivan of the American Petroleum Institute, and Thomas B. Sleeman of Union Oil Company of California. The deficiencies that remain, needless to say, are our responsibility.

J.G.M.
L.N.

 *Chapter 1*

Introduction and Summary

Popular estimates of the ability of the manufacturing sector to save energy have ranged from none at all to a great deal. Many who held the former view felt that there were rigid energy "requirements" to produce a ton of steel or a bag of cement, and thus any reduction in energy use would automatically bring with it a lower level of output. Those who held the latter view include critics who berate manufacturers for not immediately adopting the latest energy-saving technology and others who speak scornfully of industry's failure to eliminate energy waste, often citing the second law of thermodynamics.

With time a more balanced view has begun to gain credence. This has come about from studies of the ways in which energy is used in manufacturing, covering the topic from the most detailed level to the most aggregated. The results of these studies can be summarized fairly easily although by its very nature such a summary ignores a host of complications.

A definition of energy conservation is a useful first step in this discussion: we define energy conservation as a decline in energy use per unit of output. This definition accords with the general notion of an improvement in the technical efficiency of energy use. It implies nothing about motivations and refers only to objective developments. In Chapter 2 the measurement of energy use and output is discussed in detail; here we simply state that energy is measured in terms of its heat content (British thermal units) to permit the quantities of various fuels and purchased electricity to be combined into one physical measure, and output is measured in value terms (at constant prices) to reflect changes in the product mix of an industry or group of industries over time. This procedure provides a common measure that permits comparisons among industries and makes it possible to trace movements in the energy-output ratio for all manufacturing to changes in component industries.

DETERMINANTS OF ENERGY USE IN MANUFACTURING

At any time, the total amount of energy used in manufacturing depends on the level and composition of demand for manufactured products, the price of energy, the quantities of capital equipment available for use in production, and the level of technology. Changes in any of these categories affect the quantity of energy consumed. If we consider each of these determinants one at a time (holding all other determinants constant), we see the following relationships:

- When the level of demand rises and more goods are produced, total energy use invariably rises as well. The production of every kind of manufactured good requires some energy and if the composition of demand (product mix) and the energy-output ratio are unchanged, a greater level of output entails greater energy use. During a business recovery period total energy use usually increases, and during a recession it often falls (or at least grows more slowly).
- When the composition of demand changes toward more energy-intensive products, energy use rises and vice versa. This product mix effect reveals itself in many industries and frequently adds to or detracts from other influences. In the absence of strong offsetting forces, increases in energy prices that result in relative price increases of energy-intensive goods induce a change in the product mix *away* from the energy-intensive goods. There is evidence that this has been occurring in U.S. manufacturing.
- When the price of energy increases, other inputs are substituted, and energy use declines (again holding other determinants constant). There is often more than one process for producing a good, and the process that uses less energy becomes more attractive when the price of energy rises. In addition, when energy becomes high in cost, more attention is paid to avoiding energy "waste" through careful housekeeping procedures.
- When the quantity of capital equipment is increased, energy use may rise or fall. An example of the former effect is a change from

a labor-intensive process to a process that uses mechanical power; that power is derived from energy use. An example of the latter is the substitution of capital for energy, which occurs, for example, when a building, steam pipe, furnace, or kiln is insulated.

- When technology improves, the same amount of output can be produced with smaller amounts of some inputs. Energy is often one of the inputs saved when this occurs. But energy use may rise with technological change, such as a new production process that may economize on other inputs (labor, for example) while using more energy. In a period when energy prices have recently risen sharply and are continuing to rise, the new technology put into use is very likely to be energy saving.

The difficulty with the preceding abstraction is that these events do not happen one at a time, and misinterpretation of causal relationships is common. For example, the price of energy rose sharply in 1975, but the level of aggregate demand simultaneously fell.[a] Then in 1976 energy prices rose further while aggregate demand rose. Manufacturing energy use fell in 1975 but rose in 1976, leading to the conclusion by some observers that the price rise had had no effect. The dominant influence in 1975 and 1976 was the level of demand, which led to reductions in both output and energy use in 1975 and increases in both in 1976. The price effect was not absent, however, for although manufacturing output was slightly higher in 1976 than in 1974, manufacturing energy use was 4 percent lower.

Another example is the growth over long periods of both the stock of capital equipment in manufacturing and the volume of energy use. This has led some observers to conclude that energy use and investment in capital goods are complementary—that is, "you can't have one without the other." Studies of specific industries reveal many instances, however, when a substantial new investment resulted in less energy use; such instances are becoming more frequent following the abrupt increases in energy prices of recent years.

In this book we aim to unravel the several effects that have operated during the last ten years on manufacturing energy use. We seek to measure the separate impacts of technical change, investment in new plant and equipment, changes in product mix, cyclical fluctuations, and the host of other factors that have impinged on manufacturing over the period so that we may present a clearer picture of what is occurring in energy conservation in this important sector of the economy.

[a]Energy prices rose in 1973 and 1974 as well, of course, but that is not germane to the example.

HOW BIG IS MANUFACTURING ENERGY USE?

This seemingly simple question can be answered in a number of ways. The most straightforward answer is to speak of energy purchased for heat and power in manufacturing. This totaled 18.1 quadrillion (quad) Btu in 1974, out of an estimated total U.S. consumption of 72.7 quad, or 25 percent (Table 1–1). But the "purchased" and "heat and power" qualifiers omit a great deal. Another 4.4 quad (6 percent) was used as raw materials and incorporated in such products as chemicals, and in lubricants, waxes, asphalt, and so on, produced by the petroleum refining industry.

Another 3.8 quad (5 percent) was used for heat and power from energy-bearing materials that were originally purchased as raw materials but which yield "captive" energy consumption; examples are metallurgical coal in steel and still gas in petroleum refining. These two additions bring the manufacturing total to 26.3 quad, or 36 percent of the U.S. total. In addition, the U.S. total includes only energy from conventional sources, while the paper industry draws about 40 percent of its energy for heat and power from wood scraps and chemical wastes (black liquors). This amounted to almost 1 quad in 1974.

The preceding discussion gives some idea of the complexity of meaningful measures of energy use and the great diversity among industries in the sources and uses of energy-bearing materials. It also indicates that more than one-third of our coal, oil, natural gas, hydropower, and nuclear power is consumed in some way in the manu-

Table 1–1. Energy Use in Manufacturing, 1974

		Quadrillion Btu	Percent of U.S. Total
Purchased for heat and power		18.11	24.9
Feedstocks for chemicals		2.48	3.4
Other petroleum products (lubricants, etc.)		1.90	2.6
Coking coal		2.23	3.1
Captive energy in petroleum refining		1.60	2.2
	Sum	26.32	36.2
(U.S. total, all sectors)		(72.67)	(100.0)

Source: U.S. Bureau of the Census, *Annual Survey of Manufactures 1974, Fuels and Electric Energy Consumed*; U.S. Bureau of Mines, *Mineral Industry Surveys*, "Crude Petroleum, Petroleum Products, and Natural Gas Liquids," December 1976; authors' estimates.

facturing sector. Developments in this sector are therefore important in determining the success or failure of the nation's efforts to conserve energy.

METHOD OF ANALYSIS

The two principal aspects of our study are the time periods and the industries studied. We examine the period before 1967 to gain historical perspective. An earlier study to which the authors contributed provides this background material [1]. The following nine years, 1967–76, can be treated as composed of three subperiods: 1967 to 1970 ("normal"), 1970 to 1974 ("transitional"), and 1974 to 1976 ("crisis"). Energy prices were generally stable (falling relative to the overall price level) from 1967 to 1970, as had been generally true during the preceding twenty years from 1947 to 1967 (Figure 1–1). The term "normal" fits this period rather well with regard to energy markets.

During the three years following 1970, we witnessed an upturn in energy prices that now seems modest, but at that time it constituted a distinct break in the historical trend. Interruptions in supplies of natural gas and gasoline also occurred, causing much concern to the manufacturing firms affected and to many motorists, and these shortages led to a series of studies of the "energy problem."

In October 1973, the Arab members of Organization of Petroleum Exporting Countries (OPEC) declared an embargo on petroleum exports to the United States. Petroleum and other energy prices then began to rise sharply in the United States, and the term "energy crisis" has come to be applied increasingly to the subsequent period.

Largely as a result of data availability, we combine the first two subperiods into one for most industries, and we compare the 1967–74 period to the 1974–76 period.[b] The first seven years are thus our yardstick for the normal period, while the following two years are considered as the crisis period.

For industry coverage, we first briefly examine total manufacturing and the five major industry groups (two-digit SIC)[c] that are the largest energy users in manufacturing. We then study the eight largest

---

[b]The U.S. Bureau of the Census conducted surveys of energy use in manufacturing for the years 1967, 1971, 1974, 1975, and 1976. Information from these surveys are a major data source for this study. The 1971 data, however, are of doubtful validity, hence our choice of years.

[c]SIC refers to the Standard Industrial Classification. "Paper and allied products" is a major industry group (SIC 26), while "paperboard mills" (SIC 2631) is an industry according to this widely used system.

Figure 1–1. Energy Prices, 1947–76

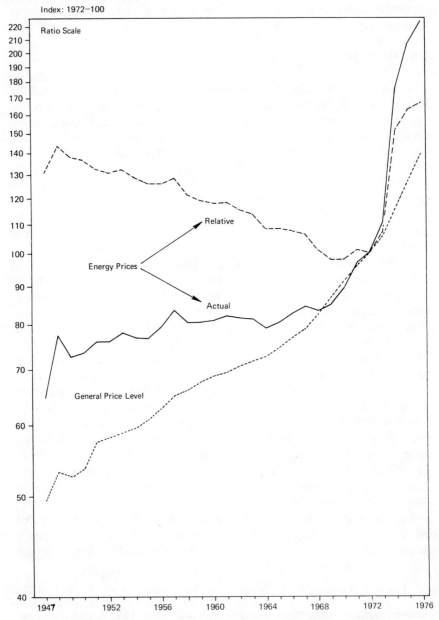

Source: (a) Energy prices, actual, from Wholesale Price Index for Fuels and Power, U.S. Bureau of Labor Statistics; (b) general price level, Implicit Price Deflator for Gross National Product, U.S. Bureau of Economic Analysis; (c) energy prices, relative, derived by dividing (a) by (b).

energy-using industries (four-digit SIC) in more detail. These include two paper industries, two chemical industries, petroleum refining, hydraulic cement, basic steel, and primary aluminum. This approach was dictated by two considerations: namely, to examine very specific developments in product mix, cyclical response, and plant turnover, and to seek the diversity in response to external pressures typical of different industries and thus avoid the pitfalls of premature generalization.

Chapters 3 to 5 report these case studies and present much statistical detail. In this chapter we offer some of the highlights of the analyses and extract findings of broad scope. Chapter 2 is devoted to an examination of some conceptual issues and includes a sketch of energy use in the economy so as to put the entire study in perspective.

POST-EMBARGO DEVELOPMENTS

Energy conservation in all manufacturing accelerated from 1974 to 1976 compared with the preceding seven years, or with the 1947 to 1967 period. The energy-output ratio (purchased energy only) fell at an average rate of 2.1 percent per year from 1974 to 1976, at a 1.2 percent rate from 1967 to 1974, and at a 1.5 percent rate from 1947 to 1967. Improvement was very uneven, however, as only two of the five largest energy-using groups showed gains (petroleum and coal products; and stone, clay, and glass products).

An analysis of changes in the composition of manufacturing output reveals that the acceleration in energy conservation since the embargo has been the result primarily of shifts away from the large energy-using industries. That is, a decline in importance of the eight industries covered in our case studies was the principal cause of the improvement in energy saving (see Chapter 2).

Within the major industry groups there is much diversity in energy saving. A brief summary of our case studies points up their value in identifying specific developments that affected energy conservation in recent years.

Papermills and Paperboard Mills

These industries, considered together for statistical purposes, experienced a *rise* in the energy-output ratio from 1974 to 1976. This development, which followed an improvement in energy conservation in the 1967−74 period, resulted from a number of negative factors: from 1974 to 1976, capacity utilization declined, investment in new plant and equipment slowed, the output mix changed toward more energy-intensive products, and more energy was utilized for pollution abatement. In the absence of these factors, energy savings would probably have accelerated. These industries obtain more than 40 percent of their energy from wood sources, and the proportion is growing over time.

Industrial Organic Chemicals, n.e.c.[d]

The pattern in this industry is similar to that of the paper industries: an acceleration in energy conservation during the 1967–74 period, when compared with earlier years, followed by a rise in the energy-output ratio from 1974 to 1976. Cyclical effects, leading to a decline in capacity utilization, are the main reason for the worsening.

Industrial Inorganic Chemicals, n.e.c.

In this industry the pattern is also repeated, but the reasons are more complex. The industry includes three uranium diffusion plants, which are very energy intensive. Output of these three plants rose rapidly from 1974 to 1976 while output in the rest of the industry rose only slightly. This change in output mix caused the overall energy-output ratio to rise. When the energy use and output of these plants are deducted from the industry totals, a steady acceleration in energy conservation is revealed for the remainder of the industry. This improvement is partly the result of a change in output mix away from alumina, an energy-intensive product.

Petroleum Refining

Very impressive gains in energy conservation were obtained in this industry from 1974 to 1976, which followed a period of no improvement from 1971 to 1974. A combination of governmental regulations caused the period of stagnation in the industry, including energy use for pollution abatement, price control, which led to operating inefficiencies, regulation-induced changes in product mix, and changes to an input mix (high-sulfur crude) that required more energy to process. The effects of these negative factors had been largely experienced by 1974, and the energy-output ratio dropped sharply thereafter.

Hydraulic Cement

Slower growth after 1967 reduced the rate of energy conservation in this industry substantially. Construction activity in the United

---

[d]The abbreviation n.e.c. stands for "not elsewhere classified" in the Standard Industrial Classification.

States slowed after 1967, leading to a slow rate of expansion in the cement industry. Annual data on energy use and output reveal a six-year period of no overall improvement in energy conservation (1967–73), followed by a gain during the next three years (1973–76) but a rate slower than that of the 1947–67 period. Substantial savings in this industry can be won only by the installation of new plant and equipment, and demand for the industry's products has not facilitated this process during the last ten years.

Blast Furnaces and Steel Mills (Basic Steel)
Energy savings were nil in this industry from 1974 to 1976 as a result of a sharp decline in capacity utilization. Basic steel is one of the two largest energy-using industries in manufacturing, and thus developments in the industry have strong effects on aggregate energy savings. In addition to the decline in output, a rise in energy use for pollution control and inefficiencies caused by interruptions in natural gas supplies contributed to the halt in energy savings, which had averaged about 1 percent per year over a long period. A renewal of output growth could bring substantial energy savings in the industry, however, as a result of increasing scrap use and the continuing introduction of more energy-efficient processes.

Primary Aluminum
A substantial acceleration in energy saving occurred in this industry from 1974 to 1976. This development followed a seven-year period (1967–74) of slow improvement, which was primarily caused by a significant slowing of output growth and relatively little expansion in capacity. Energy use for pollution control, comparatively large in this industry, contributed to the slowdown. The 1974–76 acceleration was brought about mainly by the short-term closing of the least efficient facilities during a sharp cyclical downturn. A longer term trend toward extensive recycling also aided the improvement in energy savings.

IMPLICATIONS OF THE CASE STUDIES

Analysis of the eight industries revealed a number of recurring themes, which are summarized below roughly in order of importance.

The Consequences of Slow Economic Growth
During the last ten years, growth of the economy has been slower than in the preceding two decades, partly as a result of two reces-

sions. This has led to a slower rate of investment in plant and equipment, and such investment is the vehicle by which the greatest gains in energy saving are obtained. Low rates of investment during the 1971–76 period are evident in all the industries studied just when new plant and equipment that incorporate technology designed to reduce energy use would have been introduced. The aluminum industry provides an example of the importance of this factor.

Energy conservation in manufacturing is strongly affected by slow growth because of the long life of plant and equipment in the sector. Direct information on the magnitude and effect of this factor is available for the cement industry. Of the 444 kilns in the U.S. industry in 1974, only 18 percent had been built in the preceding ten years, while 48 percent were twenty or more years old [2]. When the unit of observation is the plant (a plant may have more than one kiln), it is found that energy use per ton of output is much higher in older than in newer plants, as shown in the following tabulation of 1973 operating data.

	Thousand Btu per Ton of Cement	
Plants Built	*Wet Process*	*Dry Process*
After 1963	7,315	6,642
1934 to 1963	7,906	6,792
Before 1934	9,131	7,922

The long life of capital stock in these industries means that only a small proportion is replaced each year because of wear and tear or technical obsolescence. Expansion, therefore, accounts for most of the change in the stock, and as illustrated by the preceding tabulation, this can bring substantial gains in energy saving through the incorporation of new, energy-saving technology.

Cyclical Effects

The deep recession of 1974–75 reduced output and energy use in all the industries studied. In most of these industries, energy use did not decline as much as output, and thus the energy-output ratio rose. This relationship generally results from inefficiencies associated with low rates of operations, which can be summarized by the rate of capacity utilization. Aside from abnormal periods of "overdraft" on existing plant and equipment, a decline in capacity utilization is associated with a slower rate of decline, or absolute rise, in the energy-output ratio. Papermills and paperboard mills provide the most clear-cut example of this pattern; it is found in several other industries as well.

An exception to this effect is found in industries where large sections of plants, or entire plants, can be shut down if there is a fall in demand and it is feasible to shut down the least efficient facilities. Such a situation requires a homogeneous product and a comparatively small number of firms in the industry, both of which characterize the primary aluminum industry.

Government Regulation

There are several ways in which government regulation affects energy conservation. While many of the regulatory programs have desirable objectives, such as reducing pollution or improving the safety and health of workers, they often accomplish these objectives only at the cost of greater energy use per unit of output. It is an unfortunate fact that the timing of the energy crisis coincides quite closely with heightened concern, as expressed in legislation, for the environment and for workers' well-being. Attempts to conserve energy are hampered by requirements to reduce pollution and to improve working conditions.

In the petroleum refining industry, it is possible to identify clearly the impact on energy use of a variety of regulations. In other industries, we have only attempted to assess the impact of pollution control: here energy is used in the treatment of air and water leaving factories in order to remove pollutants. For this study, historical data on energy use for pollution control—actual figures, year by year—are needed in order to measure the impact of the pollution regulations on energy use. Most of the studies of this subject do not contain historical data, but instead they provide projections of energy use that will be required at some future date to meet a specified set of regulations.

In the absence of historical data, we have prepared estimates of energy use for pollution control in some of the industries studied and reviewed independent estimates for others. The durable goods industries, cement, basic steel, and primary aluminum, are near the upper end of the range of estimates, using between 2 percent and 3 percent of total energy for pollution control. Most of this has occurred since 1970, and a great deal of it in 1975 and 1976, as a consequence of environmental regulations. Therefore, energy conservation has been less than in the absence of these regulations, but the differences in the annual rates of energy conservation due to pollution control have been small.

A reading of the technical literature suggests that much of the energy used to comply with occupational safety and health regulations (OSHA) has been in connection with ventilation, which is powered by electricity. If this impression is correct, the method of estimation used for pollution abatement, based on growth of electricity use per unit of product, has included the OSHA effect in the totals given.

One final aspect of this problem is the diversion of investment funds from replacement and expansion to purposes required by regulations. It would require a substantial effort to make a careful estimate of the magnitude of this effect, and the problem is well beyond the scope of this study. We shall confine our discussion, therefore, to two observations. The first is that anything that slows expansion or replacement, or investment in retrofit equipment designed to save energy, hinders energy conservation in a single industry. The second observation is that there are complex interindustry effects that can change the overall product mix in manufacturing, thus producing unexpected results. Capital diversion, whatever its magnitude, could either raise or lower the overall rate of energy conservation in manufacturing.

VOLUNTARY INDUSTRIAL ENERGY CONSERVATION PROGRAM (VIECP)

There are two aspects to this program, presently administered by the U.S. Department of Energy (DOE) under the provisions of the Energy Policy and Conservation Act of 1975, relevant for this study. Goals for energy conservation by the year 1980, which vary by major industry group, were set by the federal government following a series of hearings and consultants' reports. To the extent that efforts to achieve these goals result in greater energy savings than would otherwise occur, the VIECP program is an additional determinant of energy conservation. Industry spokesmen have assured us that this is in fact the case. We have not succeeded in devising a procedure to measure this factor, however, and must therefore confine ourselves to a recognition of its possible effect.

The VIECP is also a data source. A number of trade associations in manufacturing collect data semiannually from member firms and furnish summaries to DOE. These data summaries are used to monitor progress toward the energy conservation goals; information drawn from these summaries is published semiannually by the federal government.

We have examined the reports of several trade associations; some of these are discussed in Chapters 3 to 5. In nearly all cases, we have concluded that the information in the reports is not applicable to this study. Our conclusion is based on four characteristics of the VIECP reports: the question they seek to answer, the industrial coverage, the proportion of the industry covered, and the time period covered.

The VIECP reports provide information that is relevant to the question: "If product mix, input mix, and fuel mix were constant over time, what would be the rate of energy conservation?" The degree to which corrections for these factors are made varies among the reporting trade associations, but the objective of the reporting systems we have examined appears to be that stated. Additional corrections for the effects of pollution control and worker safety and health are sometimes appended to the reports with the same objective, namely, to derive a rate of technical change in energy use. In this study, a quite different question is addressed: "What was the actual rate of energy conservation and how can it be explained?" Information collected under the VIECP has frequently been extensively adjusted beforehand and is not usable in attempting to answer the question that concerns us.

Conservation goals set under the VIECP are for major industry groups (two-digit SIC). Since the reporting units are firms, rather than plants, their reports often cover a number of industries (four-digit SIC) within a major industry group. As stated earlier, we feel strongly that a much better understanding of energy conservation can be gained from the study of narrowly defined industries than of broader, more heterogeneous groupings. This consideration makes some VIECP reports inappropriate for our study. A related problem is the proportion of the industry covered. Data collected by the U.S. Census Bureau covers nearly all the plants in an industry with the exception of the very smallest. Several trade associations reporting under the VIECP, however, have much more restricted coverage.

Finally, we believe that energy-use data for an industry that extend over several years are valuable in analyzing energy conservation. A long series permits the identification of long-term developments and provides a yardstick against which to measure recent performance. The VIECP reports cover only the period from 1972 to date; even these short series are discontinuous for some industries because coverage of firms, reporting methods, and so on, have changed.

NOTES TO CHAPTER 1

1. J.G. Myers, et al., *Energy Consumption in Manufacturing* (Cambridge, Ballinger Publishing Co., 1974).

2. Portland Cement Association, *Energy Conservation Potential in the Cement Industry*, National Technical Information Service (NTIS), June 1975, pp. 11 and 18.

Chapter 2

Background of the Study and Measurement Issues

The importance of manufacturing in total U.S. energy use was summarized in Table 1–1. In this chapter we begin with a discussion of the measurement of energy and output, continue with a discussion of factors affecting energy conservation, and the present a sketch of manufacturing energy use and its effect on the national total.

ENERGY DATA

The basic source materials on energy used in this report are the publications of the U.S. Bureau of the Census, especially the *Census of Manufactures* for 1967 and the *Annual Survey of Manufactures* for 1974, 1975, and 1976. Data on energy use were also collected in conjunction with the *1972 Census of Manufactures* (hereafter referred to as the 1972 census for economy of expression) for the year 1971. In the 1967 census and the 1974–76 surveys, the energy data are for the same year. Census Bureau data thus provide us with five observations for the ten years, including annual observations for later years.

One reason for relying mainly on census data is comprehensiveness. Energy use is collected for every plant (except the very smallest) in every industry in manufacturing, using the same definitions and collection procedure. A large degree of comparability is thus provided across industries. A second advantage is the length of the historical series, which extends back to 1947.[a] But perhaps the most

[a]The earlier years covered are 1947, 1954, 1958, and 1962.

important aspect is the wide range of complementary information that is published for the same plants, such as detailed data on products, material inputs, employment, and so on, which aid in the interpretation of the energy data.

There are limitations to the census data, however, which have led us to substitute information from other sources in some of the case studies. One is the lack of annual data prior to 1974. Annual series have several advantages for analysis. Data errors are picked up more readily, and changes in industry structure, input mix, output mix, and technology are identified more easily from yearly figures. Such data are particularly important for industries that are subject to sharp cyclical fluctuations.

The period since 1967 has embraced two recessions and recoveries, and these have significant impacts on durable goods industries (Figure 2–1). We have used annual data in our case studies of the cement and petroleum refining industries (from the Bureau of Mines [BOM]) and the steel industry (from the American Iron and Steel Institute [AISI]). No annual series is available for primary aluminum for the 1967–76 period; this is unfortunate since this industry is subject to great cyclical variation. Annual data help to avoid such errors as mistaking a cyclical fluctuation for a long-term development; this is particularly important for durable goods industries such as cement, steel, and primary aluminum.

One of the advantages enjoyed by the Census Bureau in data collection is their steadfast guarantee of nondisclosure. This encourages companies to report sensitive data. There is a negative aspect, however, in that some basic information is not published by the Census Bureau in order to avoid disclosure of the activities of individual plants. An example may be seen in the primary aluminum industry, for which almost no information on the use of individual fuels is published. Alternative information sources, such as BOM and AISI, are less restricted in the detail they publish.

"CAPTIVE" ENERGY

The petroleum refining industry derives a large proportion of its heat and power from so-called "captive consumption" of energy that is a by-product or coproduct of crude petroleum used as a raw material (Table 1–1). These captive sources are refinery (still) gas, petroleum coke, and residual fuel oil; they are not available in Census Bureau publications for 1971 or 1974–76. Similarly, metallurgical coal and its by-products or coproducts (coke and coke-oven gas) are not reported as sources of heat and power in the steel industry in Census Bureau statistics for the years 1971 and 1974–76.

Figure 2–1. Cyclical Variations in Industrial Production

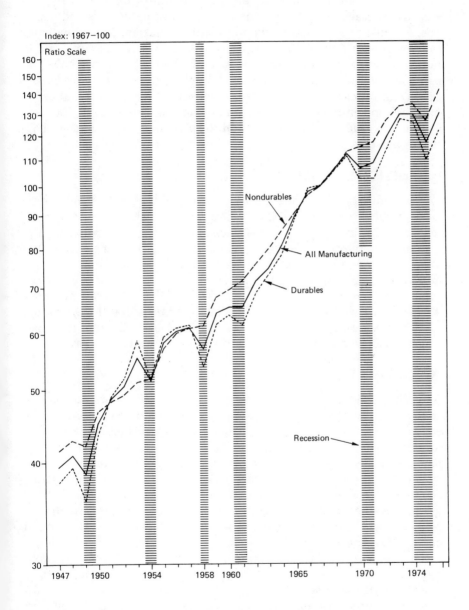

Source: Federal Reserve Board; U.S. Department of Commerce.

Three other industries covered in our case studies—papermills, paperboard mills, and industrial organic chemicals, n.e.c.—also use substantial quantities of captive energy. No information on this energy use is available in the census statistics. For papermills and paperboard mills, the American Paper Institute has made estimates of captive energy from 1972 on; none is available for earlier years to our knowledge. For industrial organic chemicals, n.e.c., we have prepared estimates of captive energy for selected, large-volume chemicals. Data on captive energy used in petroleum refining are published by BOM and in steel by AISI. Although this information is discussed in the case studies, it is not used in the statistical tables in this chapter, to preserve comparability of coverage among industries.

DEFINITION OF ENERGY

For all industries, the measure of energy consumption is the energy used for heat and power. Energy-bearing materials used as raw materials in a manufacturing process, such as crude oil in petroleum refining, are not counted in energy consumption except to the extent that they result in captive consumption for which we can obtain data (as in petroleum refining). Specifically excluded from our energy consumption measure are petrochemical feedstocks used as raw materials in the chemical industry. The rationale for this is that if the product of an industry incorporates energy-bearing materials, such materials have not been used as "energy" in the manufacturing process.

"GROSS ENERGY"

A further dimension of our energy measure is the treatment of purchased electricity. Kilowatt-hours (kwh) of purchased electricity are converted to Btu by the average fuel use in electric utilities per kwh produced (usually referred to as the "heat rate"). This means that purchased electricity is valued by the energy used to produce it rather than the heat content obtainable from it. There is a substantial difference between the two measures. For example, electric utilities used an average of 10,369 Btu of fuel to produce a kwh of electricity in 1976, whereas the heat content of a kwh is only 3,412—about one-third of the 1976 fuel requirement.

The purpose of this treatment of electricity is twofold. First, our method (sometimes termed "going back to the powerhouse") provides a more complete measure of the demand by an industry for energy resources; an industry that uses a great deal of purchased electricity is drawing much more on energy supplies than would be indicated by a conversion ratio of 3,412. Second, when an industry changes the proportion of all the electricity it uses that is purchased (e.g., buying more from utilities and generating less itself from coal or oil), the energy use of the industry does not rise or fall from this factor alone (except as it reflects a different heat rate in utilities from that of the industry's own generation).

One aspect of this treatment of electricity, which may seem strange at first, is that all electricity is converted at the fossil-fuel rate, including that produced from hydropower and nuclear power. The heat required to generate a kwh from nuclear power is not dissimilar from that for fossil-fuel plants [1]. For hydropower, the rationale is that the possibilities of expansion of this source are limited by environmental considerations, and thus if an additional kwh is produced, it will come from a fossil-fuel (or nuclear) plant; the fossil-fuel heat rate is therefore the added, or marginal, cost of a kwh and is the correct measure to be used.

CONVERSION TO BTU

Quantities of purchased fuels and electricity are converted to a common measure, the British thermal unit, by multiplying each quantity by a conversion factor that represents the average heat content of a unit of that source. For purchased electricity, this is the heat rate for fossil-fuel electric plants, discussed in the preceding section. The heat rate, computed by the Edison Electric Institute from data published by the Federal Power Commission, varies year by year. It fell by one-third from 1947 to the mid-1960s, but it has not changed appreciably since that time.

For individual fuels, conversion factors were taken from a variety of sources, but they do not differ appreciably from those used in an earlier work to which the authors of this book contributed [2]. All the conversion factors used in this work are given in Appendix A together with a description of some special problems that arise in converting published energy data.

The results of the conversion procedure are shown in Table 2–1. Despite the omission of captive energy, the numbers reveal the substantial fraction of energy use in all manufacturing accounted for by the eight industries we are studying. This is particularly true of the steel and petroleum refining industries, which account for 10 percent and 8 percent, respectively, of the total in 1976. In sum, the eight industries absorbed nearly one-half of the total, and the five major industry groups (of which the eight are part) absorbed 70 percent in 1976. The substantial decline in energy use from 1974 to 1975, which was only partly made up in 1976, is also noteworthy.

Table 2–1.   Gross Energy Purchased by High-Energy-Using Manufacturing Industries, Selected Years, 1967–76
(trillion Btu)

Industry or Industry Group (SIC number and title)	1967	1971	1974	1975	1976
2621 Papermills, except building paper[a]	614	713	707	661	740
2631 Paperboard mills[a]	491	518	587	530	539
2818 Industrial organic chemicals, n.e.c.[a,b]	982	1,095	1,270	1,185	1,355
2819 Industrial inorganic chemicals, n.e.c.[b]	932	816	965	1,035	1,080
2911 Petroleum refining[a]	1,477	1,769	1,703	1,492	1,446
3241 Hydraulic cement	521	519	544	484	493
3312 Blast furnaces and steel mills[a]	1,825	1,725	1,952	1,632	1,744
3334 Primary aluminum	589	596	864	680	706
All manufacturing	15,649	17,298	18,113	16,668	17,353
26 Paper and allied products	1,398	1,590	1,625	1,505	1,620
28 Chemicals and allied products	3,309	3,524	3,972	3,804	4,069
29 Petroleum and coal products	1,562	1,858	1,800	1,582	1,535
32 Stone, clay, and glass products	1,414	1,520	1,556	1,400	1,447
33 Primary metal industries	3,356	3,374	3,884	3,249	3,438

[a] Captive consumption of energy (not shown) is particularly important in these industries. See later chapters for estimates.
[b] The establishments in these industries were extensively reclassified after 1971. Figures for subsequent years are estimates by the authors; see Chapter 4 for details.

Source: *Census of Manufactures,* 1967 and 1972; *Annual Survey of Manufactures,* 1974, 1975, and 1976.

OUTPUT

The output measures used in this chapter were all derived by combining indexes of production taken from the 1972 Census of Manufactures with components of the Federal Reserve Board Index of Industrial Production (FRB index). For each industry, major industry group, and all manufacturing, the corresponding census production index, which shows the change in output from 1967 to 1972, was interpolated (for 1971) and extrapolated (for 1974−76) by the component of the FRB index that agrees most closely in industrial coverage (see Appendix B for details). We obtained a measure of real output for each industry by taking the dollar amount of value added by that industry in 1974 and carrying it forward and back by the census-FRB index.

Value added is a dollar measure, computed by the Census Bureau by subtracting the total cost of materials of an industry from the value of its production and then adjusting for inventory change. Value added is thus a measure of work performed in an industry during a year. It was not feasible to estimate directly the value added of each industry in dollars of constant purchasing power for all of the years that interest us. But the real output we have computed provides a close approximation to such an industry measure. The set of constant-dollar value-added measures for the industries and groups we are studying is presented in Table 2−2.

It is noteworthy that the output of all manufacturing was practically the same in 1976 as in 1974. Output declined in five of the eight industries we are studying over this two-year period and in two of the five major industry groups shown in Table 2−2. In total, the eight large energy-using industries and the five major industry groups of which they are part fared worse than the rest of manufacturing in the period of recession and partial recovery encompassing the years 1974 to 1976.

As mentioned above, all the output measures shown in Table 2−2 and discussed in this chapter were computed by the same procedure, are comparable with one another, and can be used to obtain residuals and other derived measures. A detailed examination of each of the census output indexes, including comparisons with other data sources (trade associations, BOM, or others), has led us to conclude that some of these indexes are inaccurate indicators of output changes from 1967 to 1972. Specifically, we have found that the census indexes for industrial organic chemicals, n.e.c., and hydraulic cement show too great a rise over the period while those for blast furnaces and steel mills and primary aluminum show too small a

Table 2-2. Value Added by Manufacture, High-Energy-Using Industries, Selected Years, 1967-76
(billion 1974 dollars)

Industry or Industry Group (SIC number and title)	1967	1971	1974	1975	1976
2621 Papermills, except building paper	3.37	3.88	4.35	3.81	4.31
2631 Paperboard mills	2.24	2.78	2.98	2.51	2.85
2818 Industrial organic chemicals, n.e.c.[a]	3.89	6.04	7.88	6.88	8.13
2819 Industrial inorganic chemicals, n.e.c.[a]	3.43	3.58	4.11	4.12	4.51
2911 Petroleum refining	7.06	7.91	8.36	8.36	8.95
3241 Hydraulic cement	1.22	1.34	1.37	1.17	1.22
3312 Blast furnaces and steel mills	15.09	13.89	17.43	13.87	15.13
3334 Primary aluminum	.96	1.08	1.33	1.05	1.15
All manufacturing	359.34	380.44	452.48	406.67	452.83
26 Paper and allied products	14.95	17.28	19.10	16.51	18.88
28 Chemicals and allied products	27.62	35.03	44.43	41.03	47.19
29 Petroleum and coal products	8.23	9.37	9.95	9.90	10.62
32 Stone, clay, and glass products	11.15	12.21	14.60	12.92	15.04
33 Primary metal industries	31.51	30.60	37.30	29.21	32.99

[a] The establishments in these industries were extensively reclassified after 1971. Figures for subsequent years are estimates by the authors; see Chapter 4 for details.

Source: *Census of Manufactures*, 1967 and 1972; *Annual Survey of Manufactures*, 1974, 1975, 1976; Federal Reserve Board.

rise. We have therefore derived an adjusted measure for organic chemicals and adopted alternative output measures for the three other industries; these are used in the case studies in subsequent chapters and described there.

For the industries where annual energy data are available, petroleum refining, steel, and cement, we use a simple physical measure of output in our case studies (barrels of petroleum, tons of steel, and tons of cement) taken from the same source as the energy data. This practice ensures a degree of comparability between the measures of energy and output, but it suffers from the failure to reflect changes in quality of output. A weighted index of output based on highly disaggregated data rises more than a count of, say, tons if the output mix shifts toward higher quality (i.e., more valuable) products and vice versa. The FRB index provides an approximation to such an index. In several of the industries we have examined, such a shift has taken place, and thus the simple measure tends to differ from the true rise in output in these industries. This is more important for long-term than for short-term comparisons, but it can produce considerable distortion in a nine-year period, such as 1967 to 1976.

HISTORICAL PERSPECTIVE

Before examining the pattern of energy conservation suggested by the figures in Tables 2–1 and 2–2, an examination of what might be called the normal pattern of energy use is desirable to gain historical perspective. The following summary is drawn from the study referred to earlier in this chapter [3].

- During the 1947–67 period (the analysis did not go beyond the latter year for most industries), energy use per unit of output fell in manufacturing; this was true for the total and for most of the thirty-odd individual industries and industry groups examined. Any subsequent development, particularly during the crisis period, must be viewed against this background of falling energy use per unit of output (measured in physical units) in a period of falling or stable real energy prices.
- The decline in the energy-output ratio in manufacturing was the net result of a number of influences, some tending to raise and others to lower the ratio.
- The dominant influence was the introduction of new technology that permitted an industry to produce a given volume of output with a smaller quantity of some inputs. Labor and energy were frequently the inputs that were economized.

- The main impetus for introducing the new technologies was general cost saving. Their introduction nearly always entailed the putting into place of new capital—that is, plant and equipment.
- Energy saving in the 1947−67 period was, therefore, generally associated with investment in new plants or with major modifications of old plants.

In view of these findings, it seems reasonable to use the long-term pattern of the 1947−67 period as a yardstick with which to compare developments after 1970. In addition, the mechanism by which past energy savings were realized—the introduction of new technology through investment in plant and equipment—furnishes a guide to understanding one type of energy conservation in the years after 1970.

ENERGY CONSERVATION

As stated in Chapter 1, our yardstick of energy conservation is reduction in energy use per unit of output. Figures on Btu per dollar of value added (in 1974 prices) are shown in Table 2−3. A summary of annual rates of change of these energy-output ratios is shown in Table 2−4.

When we compare 1974−76 (crisis) with 1967−74 (normal), we see that energy conservation clearly accelerated for the manufacturing sector as a whole from a 1.2 percent average rate of decline to a 2.1 percent rate. An acceleration occurred in only three of the nine industries shown, however, and in only two of the five major industry groups. In fact, the energy-output ratio *rose* in five of the eight industries and two of the five major industry groups. Part of this somewhat puzzling result is explained by the bottom lines of Table 2−4. In manufacturing industries other than the five major groups (considered as a group), energy conservation slowed, the average annual rate declining by one-half. In the eight industries combined, energy conservation slowed. An acceleration took place, therefore, in industries other than the eight we are studying but within the five major industry groups shown in Table 2−4. That is, part of the acceleration in energy conservation in all manufacturing was brought about by developments in industries in the chemicals, stone, clay, and glass, and so on, groups other than the eight shown in Table 2−4.

We have examined the energy-output patterns of a small number of industries in the chemical (SIC 28) and stone, clay, and glass (SIC 32) groups *other* than the eight shown in Table 2−4. The results indicate that their rate of energy conservation did indeed accelerate in the 1974–76 period, supporting the observation presented in the preceding paragraph.

The principal cause of the acceleration, however, was a change in product mix. The level of the energy-output ratio is much higher for the five major industry groups (combined) than for the remainder of manufacturing. The energy-output ratio for the eight industries (combined) chosen for case study is also much higher than for the other industries in the five industry groups (combined).

	Energy-Output Ratio *(1,000 Btu per 1974 dollar of value added)*		
	1967	*1974*	*1976*
All manufacturing	43.5	40.0	38.3
Eight industries (combined)	199.5	179.7	175.2
Five industry groups (combined)	188.1	102.4	97.1
Other industries in five groups (combined)	64.2	54.7	51.0
All other than five groups (combined)	17.3	16.1	16.0

When output expands more in industries with *high* energy-output ratios than in industries with *low* energy-output ratios, the average rises from this effect alone and vice versa. From 1967 to 1974, there was a shift in the composition of manufacturing output toward the energy-intensive industries (those with higher energy-output ratios). This shift caused the energy-output ratio of all manufacturing to decline *less* than it would have with an unchanged output mix. From 1974 to 1976, there was a shift away from the energy-intensive industries, and thus the energy-output ratio fell more than it would have without the shift.

	Percent of All Manufacturing		
	1967	*1974*	*1976*
Value added in all manufacturing	100.0	100.0	100.0
Sum of eight industries	10.4	10.6	10.2
Sum of five industry groups	26.0	27.7	27.5
Other industries in five groups	15.6	17.0	17.3
All other than five groups	74.0	72.3	72.5

A computation of the magnitude of this effect suggests that energy conservation in the 1967–74 period was *reduced* by one-third by the shift in product mix to the energy-intensive industries and

Table 2–3. Gross Energy Purchased per 1974 Dollar of Value Added, High-Energy-Using Manufacturing Industries, Selected Years, 1967–76 (1,000 Btu per 1974 dollars)

Industry or Industry Group (SIC number and title)	1967	1971	1974	1975	1976
2621 Papermills, except building paper[a]	182.5	183.8	162.6	173.7	171.6
2631 Paperboard mills[a]	219.2	186.7	196.9	211.1	189.0
2818 Industrial organic chemicals, n.e.c.[a,b]	252.4	181.4	161.2	172.3	166.7
2819 Industrial inorganic chemicals, n.e.c.[b]	271.8	228.0	234.8	251.2	239.4
2911 Petroleum refining[a]	209.2	223.6	203.6	178.4	161.5
3241 Hydraulic cement	428.0	389.0	396.1	414.5	402.9
3312 Blast furnaces and steel mills[a]	120.9	124.2	112.0	117.6	115.3
3334 Primary aluminum	610.5	549.9	649.7	645.5	613.9
All manufacturing	43.5	45.5	40.0	41.0	38.3
26 Paper and allied products	93.5	92.0	85.1	91.1	85.8
28 Chemicals and allied products	119.8	100.0	89.4	92.7	86.2
29 Petroleum and coal products	189.9	198.3	180.9	159.7	144.5
32 Stone, clay, and glass products	126.7	124.5	106.6	108.3	96.2
33 Primary metal industries	106.5	110.3	104.1	111.2	104.2

[a]Captive consumption of energy (not shown) is particularly important in these industries. See later chapters for estimates.
[b]The establishments in these industries were extensively reclassified after 1971. Figures for subsequent years are estimates by the authors; see Chapter 4 for details.

Source: Data underlying Tables 2–1 and 2–2 (this table computed from unrounded figures).

Table 2—4. Energy Conservation in High-Energy-Using Manufacturing Industries, 1967—74 and 1974—76: Average Annual Rate of Change in Gross Energy Purchased per 1974 Dollar of Value Added *(percent)*

Industry or Industry Group (SIC number and title)	1967 to 1974	1974 to 1976
2621 Paper mills, except building paper	-1.6	+2.7
2631 Paperboard mills	-1.5	-2.0
2818 Industrial organic chemicals, n.e.c.	-6.2	+1.7
2819 Industrial inorganic chemicals, n.e.c.	-2.1	+1.0
2911 Petroleum refining	-0.4	-10.9
3241 Hydraulic cement	-1.1	+0.9
3312 Blast furnaces and steel mills	-1.1	+1.5
3334 Primary aluminum	+0.9	-2.8
All manufacturing	-1.2	-2.1
26 Paper and allied products	-1.3	+0.4
28 Chemicals and allied products	-4.1	-1.8
29 Petroleum and coal products	-0.7	-10.6
32 Stone, clay, and glass products	-2.4	-5.0
33 Primary metal industries	-0.3	0.0
Sum of eight industries (four-digit SIC)	-1.5	-1.3
Sum of five industry groups (two-digit SIC)	-2.0	-2.6
All other than five industry groups	-1.0	-0.5

Source: Table 2-3.

that it was *increased* by one-third in the 1974—76 period. The 1974—76 period was characterized by a sharp rise in energy prices and by a decline in total manufacturing output, as we have observed. The energy price increase falls more heavily on the energy-intensive industries, causing them to lose ground compared with less energy-intensive industries; this explains, in part at least, the product mix shift in the 1974—76 period. The decline or slow growth in output experienced by the eight industries also made it difficult for them to conserve energy.

METHODS OF ACHIEVING ENERGY CONSERVATION

Energy conservation can be achieved in a number of ways. It is instructive to list some of these ways and to attempt to group them by difficulty of attainment. As stated in Chapter 1, we are not restricting the term "energy conservation" to the results of conscious acts toward that end. Indeed we consider that to be an impossible task that would require identification of motives. Instead we are examining objective developments that result in using less energy per unit of output.

We are also examining energy conservation within a given regulatory framework. This leads us to classify methods of achieving energy conservation according to the amount of capital investment needed to carry out each method from the lowest to the highest. If the regulatory framework were changed, the ranking of difficulty might change, of course.

Process Monitoring

This includes maintenance to improve the efficiency of energy use as well as modification of operations. Such methods generally require the application of labor and management but only minor investment expenditures. Yet some technical experts maintain that there are very substantial energy savings to be gained in this way.

Changes in Output Mix

Products vary substantially in their energy "requirements" per pound or other physical measure. Shifts among products can either lower energy use or raise it. Such shifts can come about from the differential impact of energy price increases, which are then reflected in product prices by diverting product demand to the less energy-intensive substitutes.[b] Changes in output mix can occur among as well as within industries, of course.

Retrofit

Existing capital equipment can be modified to produce energy savings. An obvious example is insulation. Methods of recovery and utilization of so-called waste heat also come under this heading. Examples of such methods are recovery of heat from kilns or furnaces to heat raw materials. Retrofit procedures can be very expensive in capital outlays. Expenditures for retrofit are the bulk of what are generally referred to as energy conservation investments in proposals for tax write-offs or other subsidies.

New Plant and Equipment

The greatest changes in energy use per unit of output can be obtained through investment in entirely new capital that has been

[b]It should be noted that we are discussing here the energy consumed during production within manufacturing industries, not the energy consumed later when using these products. For example, it is the energy consumed in producing an air conditioner that concerns us here, not the energy consumed by the air conditioner in cooling a room.

designed with energy savings in mind. Savings can be achieved in several ways with such new capital: (1) In some industries alternative production processes are presently available that have differing capital-energy ratios (aluminum smelting and chlorine production are examples); here a movement to a more capital-intensive process reduces energy use per unit of product. (2) The design of new factories and machinery can be modified to reduce energy use; this can be accomplished through reduction in heat radiation, reducing time intervals between stages in manufacture, or other improvements in efficiency. (3) Processes can be integrated in new plants, such as locating pulping operations adjacent to paper- or paperboard-making facilities, processing hot metal direct from smelters, and so on. (4) Entirely new production processes can be introduced that use less energy; these require research and development expenditures, which are another form of investment related to energy conservation.

ENERGY INTENSIVENESS AND ENERGY USE

The level of the energy-output ratio (Btu per dollar of value added in constant prices) of a manufacturing industry provides a measure of the energy intensiveness of that industry. The more Btu per dollar of product, the more energy intensive the industry.[c] In Table 2-5, the nineteen manufacturing industries are shown that used the most (purchased) energy in 1974, as well as the largest five groups and total manufacturing.

Restricting the energy measure to energy "purchased for heat and power," as in this chapter, affects the comparison. If the Btu content of coking coal were added to the steel industry and captive consumption of energy were added to the petroleum refining industry, the measure of energy intensiveness of each of these industries would approximately double, raising their ranks. Other industries, particularly industrial organic chemicals, n.e.c., would also be affected, but to a lesser degree. The information necessary to adjust these industry totals to a broader concept that would make them comparable (as they are now on the narrow concept) does not exist, however, because the alternative data sources used for the case studies described in later chapters use different definitions and have different coverage.

---

[c]An alternative measure would be expenditure on energy as a percentage of product value. This measure would provide a similar ranking, but it would differ somewhat because the cost per Btu varies across industries. See the discussion later in this chapter.

Table 2–5. Gross Energy Purchased Compared with Value Added,
High-Energy-Using Manufacturing Industries, 1974

Industry or Industry Group (SIC number and title)	Gross Energy (trillion Btu)	Value Added (million 1974 $)
	(1)	(2)
3312 Blast furnaces and steel mills[a]	1,952	17,426
2911 Petroleum refining[a]	1,703	8,364
2869 Industrial organic chemicals, n.e.c.[a,b]	1,246	7,660
3334 Primary aluminum	864	1,330
2621 Papermills, except building paper[a]	707	4,347
2819 Industrial inorganic chemicals, n.e.c.[b]	688	2,904
2631 Paperboard mills[a]	587	2,980
3241 Hydraulic cement	544	1,374
2873 Nitrogenous fertilizers[b]	298	1,141
2812 Alkalies and chlorine	264	698
2821 Plastics materials and resins	245	3,640
2824 Organic fibers, noncellulosic	201	2,411
2865 Cyclic crudes and intermediates	197	1,465
3714 Motor vehicle parts and accessories	192	10,250
3079 Miscellaneous plastics products	191	8,107
2813 Industrial gases	170	544
3221 Glass containers	169	1,498
3321 Gray iron foundries	161	3,035
3711 Motor vehicles and car bodies	159	10,849
All manufacturing	18,113	452,478
26 Paper and allied products	1,625	19,096
28 Chemicals and allied products	3,972	44,432
29 Petroleum and coal products	1,800	9,951
32 Stone, clay, and glass products	1,556	14,600
33 Primary metal industries	3,884	37,297

[a]Captive consumption of energy (not shown) is particularly important in these industries. See later chapters for estimates. Inclusion of captive consumption would modify the rankings somewhat.

[b]Nitrogenous fertilizers (2873), an industry first tabulated in 1972, was formerly divided among industrial organic chemicals, n.e.c. (2869, formerly 2818), industrial inorganic chemicals, n.e.c. (2819), and fertilizers (formerly 2871). Other reclassifications were also made in chemical industries in 1972.

Source: *Annual Survey of Manufactures*, 1974.

Despite this limitation, an important characteristic of manufacturing energy use can be observed from the data in Table 2–5; the high-energy-using industries are also energy intensive. All the five major industry groups shown in the table and sixteen of the nineteen

Table 2—5. continued

Ratio Gross Energy to Value Added (1) ÷ (2)	*Rank According to:*	
(1000 Btu/1974 $)	*Gross Energy (1)*	*Energy Intensiveness (3)*
(3)	(4)	(5)
112	1	13
204	2	7
163	3	9
650	4	1
163	5	9
237	6	6
197	7	8
396	8	2
261	9	5
378	10	3
67	11	15
83	12	14
134	13	11
19	14	18
24	15	17
313	16	4
113	17	12
53	18	16
15	19	19
40		
85		
89		
181		
107		
104		

industries shown are more energy intensive than total manufacturing. Furthermore, fourteen of the nineteen industries are more than twice as energy intensive as the total.

ENERGY SOURCES

Three-fourths of energy purchased by manufacturing industries comes from two sources, electricity (converted on a gross basis) and natural gas, in similar proportions (Table 2—6). Purchased electricity and fuel oil grew rapidly in importance over the 1967—76 period while coal declined. Natural gas rose from 1967 to 1971 and then fell back in importance from 1971 to 1976.

**Table 2-6. Purchased Energy Used by High-Energy-Using Groups
Distributed by Source, Selected Years, 1967-76** *(percentage distribution)*

Industry Group (SIC number and title)	Year	Fuels and Electricity	Purchased Electricity
		(1)	(2)
All manufacturing			
	1976	100	38
	1975	100	37
	1974	100	36
	1971	100	31
	1967	100	28
26 Paper and allied products			
	1976	100	28
	1975	100	27
	1974	100	26
	1971	100	23
	1967	100	19
28 Chemicals and allied products			
	1976	100	36
	1975	100	35
	1974	100	33
	1971	100	30
	1967	100	30
29 Petroleum and coal products			
	1976	100	19
	1975	100	17
	1974	100	16
	1971	100	13
	1967	100	12
32 Stone, clay, and glass products			
	1976	100	21
	1975	100	21
	1974	100	19
	1971	100	17
	1967	100	14
33 Primary metal industries			
	1976	100	45
	1975	100	44
	1974	100	44
	1971	100	38
	1967	100	34
All other groups			
	1976	100	48
	1975	100	47
	1974	100	46
	1971	100	41
	1967	100	36

[a]Includes coke.

[b] Z represents a figure less than 0.5.

Source: *Census of Manufactures; Annual Survey of Manufactures.*

Table 2–6. continued

All Fuels	Fuel Oil	Natural Gas	Coal and Coke	Other Fuel
(3)	(4)	(5)	(6)	(7)
62	11	35	9	6
63	10	36	9	8
64	10	37	9	9
69	9	39	11	10
72	7	35	15	15
72	31	23	13	4[a]
73	30	25	13	5
74	28	27	13	5
77	25	31	15	5
81	18	26	24	12
64	8	42	8	6
65	7	42	9	7
67	7	44	9	7
70	6	42	14	9
70	4	38	16	12
81	6	72	Z[b]	3
83	5	72	Z	5
84	5	68	1	11
87	5	74	Z	8
88	4	74	1	8
79	10	42	20	7
79	9	42	19	9
81	8	45	16	11
83	8	48	17	10
86	5	46	22	13
55	9	28	15	3
56	9	29	13	5
56	8	29	13	5
62	7	34	17	5
66	7	35	18	6
52	11	26	5	10
53	10	26	5	11
54	9	28	5	12
59	9	27	7	16
64	7	19	11	27

The pattern just described applies with minor variation to most of the high-energy-using industries (Table 2–7). Purchased electricity grew in importance in all of these industries, fuel oil rose in six of them, and coal use fell in at least five, although there was some movement back to coal in 1976.

Wide differences existed in the average prices paid for energy during the 1967−76 period (Tables 2−8 and 2−9). The differences were substantial among different energy sources in the same industry or group, among industries for the same energy source, and among rates of price increases for the different energy sources.

Purchased electricity[d] was a more expensive source than fuels in general, while natural gas was the cheapest source. As would be expected, fuel oil rose the fastest over the period and was the most expensive source in 1974 and 1975 for every industry and group.

Electricity was purchased by the primary aluminum industry at prices less than half those paid by other industries. There are valid reasons for this difference, however. It is much cheaper to supply electricity to the primary aluminum industry than the average industry because primary aluminum uses very large volumes on a steady, round-the-clock, year-round basis. This steady load can be supplied at much lower cost per kwh than a fluctuating demand. In addition, many of the transmission lines and all transformation equipment are provided by the primary aluminum industry, so that the price paid by the industry is for a good at an earlier stage of production than is true for most industries.

The trends and patterns revealed in these tables help to explain many of the developments we shall examine in subsequent chapters. They fail to reveal one important aspect, however, which is supply interruptions. These occurred in natural gas, electricity, and fuel oil at various times and places, but their impacts often do not show in price statistics, such as those of Tables 2−8 and 2−9, because of price controls. The reduction in the importance of natural gas as an energy source shown in Tables 2−6 and 2−7 is the result of such supply interruptions (or the threat of them), but it is not reflected in the price statistics.

Another dimension to this question may be drawn from an examination of the distribution of each fuel by purchasing industry in 1974 (Table 2−10). Although the table omits captive consumption of energy, the steel industry is still by far the largest buyer of coal and coke, followed by cement. Steel is also the largest buyer of fuel oil, followed closely by paperboard mills and papermills. Natural gas purchases are centered in petroleum refining and industrial organic chemicals, n.e.c. Finally, the largest users of purchased electricity are primary aluminum and industrial inorganic chemicals, n.e.c.

[d]Electric energy is measured in gross energy inputs to generating stations. Price per delivered electric Btu is much higher.

Some of these patterns of energy use by source are dictated by the dominant production process. Examples are the use of electricity in primary aluminum and industrial inorganic chemicals, n.e.c. (which includes gaseous diffusion plants), and of coal and coke in steel. Other source patterns have resulted from price considerations—these may be easier to modify, but they often require new investments.

ENERGY USE IN THE TOTAL ECONOMY—
A REFERENCE POINT

We have defined energy conservation as using fewer Btu to produce a unit of product. This definition is correct in our judgment, but it is not directly measurable from existing statistics for the total economy. The reason lies in the conventions of national income accounting, which exclude production in the household. Therefore, energy enters into the gross national product (GNP) as both a producer's good, as in manufacturing, and a consumer's good. But it is not the electricity, gasoline, and so forth, that the consumer wants but rather the services they provide—light, heat, transportation, and so on. To the extent that these services are produced in the household, they are not counted in the GNP; instead, the energy used in their production is included.

If the ultimate services consumed, wherever produced, were counted in the GNP instead of the energy used to produce them, changes in the ratio of total energy use to GNP (in constant prices) would provide a true measure of energy conservation in our sense. If this ratio fell over time, it would indicate true conservation, or the converse. With GNP as presently measured, however, a decline in the ratio of energy to GNP reflects the net effect of changes in the sector in which production takes place (i.e., in the household or in business) and energy use per unit of output. For example, a reduction in gasoline consumption in the household sector that resulted from less fuel use per passenger mile would be indistinguishable from a reduction that resulted only from a cutback in miles driven. Yet the former represents energy conservation, according to our definition, while the latter does not.

ENERGY AND GNP IN THE POST-WAR PERIOD

With this consideration in mind, we now turn to an examination of the energy-GNP ratio (Btu per dollar of GNP in constant prices). From 1947 to the mid-1950s, this ratio fell sharply—14 percent in

Table 2-7. Purchased Energy Used by High-Energy-Using Industries Distributed by Source, Selected Years, 1967-76 *(percentage distribution)*

Industry (SIC number and title)	Year	Fuels and Electricity (1)	Purchased Electricity (2)
2621 Papermills[a]			
	1976	100	29
	1975	100	28
	1974	100	27
	1971	100	25
	1967	100	22
2631 Paperboard mills[a]			
	1976	100	20
	1975	100	19
	1974	100	18
	1971	100	14
	1967	100	11
2818 Industrial organic chemicals, n.e.c.[a,b]			
	1976	100	17
	1975	100	17
	1974	100	15
	1971	100	19
	1967	100	14
2819 Industrial inorganic chemicals, n.e.c.[b]			
	1976	100	71
	1975	100	70
	1974	100	65
	1971	100	43
	1967	100	51
2911 Petroleum refining[a]			
	1976	100	19
	1975	100	17
	1974	100	16
	1971	100	13
	1967	100	12
3241 Hydraulic cement			
	1976	100	19
	1975	100	19
	1974	100	19
	1971	100	17
	1967	100	15
3312 Blast furnaces and steel mills[a]			
	1976	100	26
	1975	100	26
	1974	100	26
	1971	100	25
	1967	100	20
3334 Primary aluminum			
	1976	100	86
	1975	100	85
	1974	100	83
	1971	100	76
	1967	100	74

[a]Captive consumption of energy (not shown) is particularly important in these industries. See later chapters for estimates.

[b] The establishments in these industries were extensively reclassified after 1971. Figures for subsequent years are based on the new classifications and therefore are not fully comparable with previous years.

All Fuels	Fuel Oil	Natural Gas	Coal and Coke	Other Fuel
(3)	(4)	(5)	(6)	(7)
71	27	22	18	4
72	27	23	18	4
73	26	24	19	3
75	23	28	21	3
78	17	22	31	8
80	40	24	13	3
81	37	29	11	4
82	36	30	12	5
86	31	35	16	5
89	21	30	26	12
83	4	63	6	9
83	4	62	7	11
85	4	65	6	9
81	2	55	12	11
86	2	55	15	15
29	4	20	3	1[c]
30	4	21	4	1
35	5	23	3	4
57	3	46	4	3
49	2	38	5	4
81	6[d]	74	Z[d,f]	1[d]
83	5	74	Z	3
84	4	69	Z	10
87	4	76	1	7
88	3	77	1	6
81	7	27	44	2[c]
81	7	34	38	3[d]
81	8	38	34	1
83	8	40	34	Z[c]
85	3	36	42	3
74	15	32	24	3
74	14	33	21	5
74	14	34	21	6
75	10	38	23	5
80	10	41	24	5
14	NA[e]	NA	NA	NA
15	NA	NA	NA	NA
17	NA	NA	NA	NA
24	Z[f]	21	3	1
26	Z	23	2	Z

[c] Includes coke. [e] NA means not available.
[d] Estimated. [f] Z represents figure less than 0.5.
Source: *Census of Manufactures; Annual Survey of Manufactures.*

Table 2—8.   Average Energy Prices Paid by High-Energy-Using Groups, Selected Years, 1967—76 *(cents per million Btu, gross energy)*

Industry (SIC number and title)	Year	Fuels and Electricity	Purchased Electricity
		(1)	(2)
All manufacturing			
	1976	159	185
	1975	139	167
	1974	108	132
	1971	60	94
	1967	49	84
26 Paper and allied products			
	1976	156	173
	1975	144	160
	1974	117	126
	1971	55	84
	1967	41	78
28 Chemicals and allied products			
	1976	133	156
	1975	113	139
	1974	86	108
	1971	47	74
	1967	36	59
29 Petroleum and coal products			
	1976	137	166
	1975	108	146
	1974	70	116
	1971	34	76
	1967	29	71
32 Stone, clay, and glass products			
	1976	149	210
	1975	131	190
	1974	100	152
	1971	55	100
	1967	45	98
33 Primary metal industries			
	1976	156	145
	1975	139	132
	1974	109	101
	1971	63	75
	1967	49	61
All other groups			
	1976	192	227
	1975	169	205
	1974	135	166
	1971	79	116
	1967	69	115

[a]Coal only.

Source: *Census of Manufactures; Annual Survey of Manufactures.*

Table 2–8. continued

All Fuels	Fuel Oil	Natural Gas	Coal and Coke
(3)	(4)	(5)	(6)
144	202	123	149
123	204	94	144
94	191	65	112
45	66	38	50
35	49	32	35
150	183	128	112[a]
139	183	96	120[a]
114	169	66	94[a]
46	57	37	45
33	38	30	31
120	202	107	106
99	207	79	106
76	193	54	89
36	67	30	41
26	46	24	26
130	200	124	106
100	191	93	100[a]
61	171	53	76
28	57	26	36
23	39	23	23
133	213	128	105
115	213	98	106
88	196	70	83
45	67	43	43
36	60	37	29
165	198	121	229
145	213	94	214
115	206	69	161
55	70	46	67
42	54	36	51
159	218	142	132
136	215	110	128
108	202	81	93
54	73	50	46
44	55	46	33

Table 2–9. Average Energy Prices Paid by High-Energy-Using Industries, Selected Years, 1967–76 *(cents per million Btu, gross energy)*

Industry (SIC number and title)	Year	Fuels and Electricity (1)	Purchased Electricity (2)
2621 Papermills			
	1976	150	159
	1975	141	151
	1974	113	117
	1971	53	76
	1967	40	70
2631 Paperboard mills			
	1976	153	158
	1975	139	141
	1974	116	117
	1971	50	81
	1967	35	71
2818 Industrial organic chemicals, n.e.c.[b]			
	1976	113	150
	1975	91	142
	1974	69	112
	1971	37	64
	1967	27	60
2819 Industrial inorganic chemicals, n.e.c.[b]			
	1976	133	137
	1975	112	114
	1974	83	83
	1971	48	68
	1967	36	46
2911 Petroleum refining			
	1976	135[c]	161
	1975	106	141
	1974	68	112
	1971	33	73
	1967	28	68
3241 Hydraulic cement			
	1976	125	196
	1975	113	171
	1974	88	132
	1971	47	89
	1967	37	85
3312 Blast furnaces and steel mills			
	1976	176	188
	1975	158	174
	1974	125	130
	1971	67	97
	1967	48	75
3334 Primary aluminum			
	1976	77	77
	1975	67	67
	1974	53	53
	1971	36	40
	1967	30	33

[a] Coal only.

[b] The establishments in these industries were extensively reclassified after 1971. Figures for subsequent years are based on the new classifications and therefore are not fully comparable with previous years.

Table 2–9. continued   *41*

All Fuels	Fuel Oil	Natural Gas	Coal and Coke
(3)	(4)	(5)	(6)
146	183	128	113[a]
138	182	99	120[a]
112	168	67	92[a]
45	57	36	44
32	38	29	31
151	179	126	111[a]
139	181	92	123[a]
115	164	62	100[a]
45	55	35	47
31	36	27	30
106	197	100	99[a]
81	203	72	97[a]
61	192	50	85
30	69	27	37
21	43	20	24
123	206	108	116[a]
106	207	84	122[a]
83	193	56	113[a]
33	68	29	46
27	49	26	27
129[c]	195[c]	124	95[a,c]
98	183	93	99[a]
59	164[d]	53	85[a]
27	53	25	36
23	33	22	22
108	175	110	97[a]
100	180	86	97[a]
78	158	59	79[a]
38	53	32	42
28	45	27	28
171	194	125	221
153	213	95	205
124	208	70	156
57	69	48	66
41	53	34	48
74	NA[e]	NA	NA
68	NA	NA	NA
53	NA	NA	NA
24	80	22	40
22	56	20	38

[c]Estimated.

[d]Fuel oil-residual only.

[e]NA means not available.

Source: *Census of Manufactures; Annual Survey of Manufactures.*

Table 2–10. Sources of Purchased Energy Distributed by Consuming Industry, 1974 *(percentage distribution)*

Industry or Industry Group (SIC number and title)	Fuels and Electricity	Purchased Electricity
All manufacturing	100.0	100.0
26 Paper and allied products	9.0	6.6
2621 Papermills, except building paper[a]	(3.9)	(3.0)
2631 Paperboard mills[a]	(3.2)	(1.6)
28 Chemicals and allied products	21.9	20.1
2818 Industrial organic chemicals, n.e.c.[a]	(6.9)	(3.0)
2819 Industrial inorganic chemicals, n.e.c.	(3.8)	(7.0)
29 Petroleum and coal products	9.9	4.4
2911 Petroleum refining[a]	(9.4)	(4.2)
32 Stone, clay, and glass products	8.6	4.6
3241 Hydraulic cement	(3.0)	(1.6)
33 Primary metal industries	21.4	26.5
3312 Blast furnaces and steel mills[a]	(10.8)	(7.9)
3334 Primary aluminum	(4.8)	(11.1)
Sum of five two-digit groups	70.9	62.3
All other manufacturing	29.1	37.7

[a]Captive consumption of energy (not shown) is particularly important in these industries. See later chapters for estimates.
[b]Includes coke.
Source: *Annual Survey of Manufactures, 1974.*

eight years, or 1.9 percent per year (Figure 2–2).[e] The ratio fluctuated during the next decade, drifting downward on balance to reach a low point in 1966. From 1966 to 1970, the energy-GNP ratio rose 8.6 percent; however, from 1970 to 1976, it has fallen each year for a total of 7 percent.

The energy figures used in the preceding summary were taken from publications of BOM [4]. Comparable figures are available from the same sources on total industrial energy use (covering the mining and manufacturing sectors combined) on an annual basis. The

[e]The four subperiods (1947–55, 1955–66, 1966–70, and 1970–76) were chosen to correspond to changes in the rate of decline of the energy-GNP ratio and the heat rate of electric utilities.

Table 2–10. continued

All Fuels	Fuel Oil	Natural Gas	Coal and Coke	Other Fuel
100.0	100.0	100.0	100.0	100.0
10.3	26.4	6.5	13.4	5.3
(4.4)	(10.5)	(2.5)	(8.3)	(1.6)
(4.2)	(12.3)	(2.6)	(4.2)	(1.7)
22.9	16.1	26.0	21.8	18.5
(9.0)	(3.0)	(12.1)	(4.9)	(7.1)
(2.0)	(1.8)	(2.4)	(1.4)	(1.5)[b]
13.0	5.1	18.2	0.6	12.5
(12.3)	(4.3)	(17.6)	(0.5)	(10.9)[b]
10.8	7.5	10.4	15.7	10.8
(3.8)	(2.6)	(3.1)	(11.3)	(0.3)[b]
18.6	18.2	17.0	32.1	12.1
(12.4)	(15.4)	(9.8)	(25.0)	(6.9)
(1.2)	(NA)	(NA)	(NA)	(NA)
75.6	73.4	78.2	83.6	59.1
24.4	26.6	21.8	16.4	40.9

industrial sector accounted for approximately the same proportion of GNP throughout the 1947–76 period (about 30 percent), and thus we can readily determine the contribution of energy conservation in the industrial sector to changes in the total energy-GNP ratio.

Over the entire twenty-nine-year period, the energy-GNP ratio fell 18 percent, and the industrial sector accounted for 86 percent of this decline. During the 1947–55 period, the industrial sector contributed just over half the decline; a sharp drop (25 percent) occurred in the heat rate of electric utilities during this period, which contributed to energy savings in both the industrial sector and the rest of the economy. From 1955 to 1966, 90 percent of the more modest decline in the energy-GNP ratio was the result of savings in the industrial sector.

The rise in the energy-GNP ratio from 1966 to 1970 was accompanied by a worsening in the industrial sector, but only one-fourth of the rise in the total was brought about by the industrial sector.

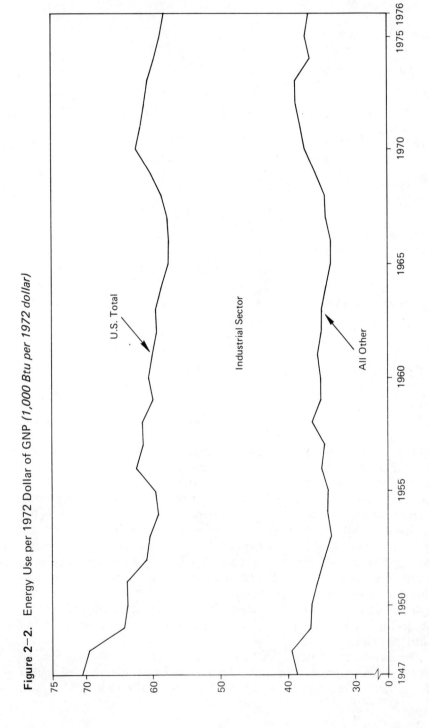

Figure 2–2. Energy Use per 1972 Dollar of GNP *(1,000 Btu per 1972 dollar)*

Source: U.S. Bureau of Mines; U.S. Department of Commerce.

Finally, 96 percent of the gains in the energy-GNP ratio since 1970 are attributable to improvement in the industrial sector. In summary, we can see that the major part of the gains in energy savings in the U.S. economy over the last thirty years are traceable directly to the industrial sector.

CAPITAL AND ENERGY CONSERVATION

We established earlier that manufacturing is a major user of energy, consuming about 35 percent of the total in 1974. Now we have also seen that manufacturing (because it makes up the preponderance of the industrial sector) contributes most of the energy savings of the nation. Our case studies clearly reveal that manufacturing industries have saved energy mainly through the introduction of new technology incorporated in new plant and equipment. That is, turnover of capital stock and expansion of the size of the stock have been the principal means by which reductions in the energy-output ratio have been achieved.

A characteristic of manufacturing industries that strongly affects their energy savings is the long life of their capital equipment. Long-lived plant and equipment mean that turnover amounts to only a small fraction of the total stock each year. So unless the manufacturing sector in general and the large energy-using industries in particular are expanding rapidly, significant savings cannot be achieved in this way. An obvious implication for government policy is that tax or other policies that promote investment will speed energy conservation.

NOTES TO CHAPTER 2

1. Bureau of Mines News Release, "Annual Energy Use Up in 1976," March 14, 1977, citing information from the U.S. Energy Research and Development Administration, sets the level for nuclear power at 10,660 Btu per kwh in 1976.

2. J.G. Myers et al., *Energy Consumption in Manufacturing* (Cambridge, Mass.: Ballinger Pub'ishing Co., 1974), p. 14 and pp. 84–5.

3. Ibid., ch. 1.

4. *Energy Perspectives II*, U.S. Department of Interior, June 1976, U.S. G.P.O., Washington, D.C. and unpublished information from the Bureau of Mines.

Chapter 3

Paper Industries

This chapter covers energy conservation in the paper industry (SIC 2621) and the paperboard industry (SIC 2631). The two industries are covered in a single chapter because the production processes used are quite similar and because the best source of current energy data, the monthly energy survey of the American Paper Institute (API),[a] combines them. These two industries also include pulp mills that are integrated with paper and paperboard mills.[b] These pulp mills account for over 80 percent of all pulp produced in the United States.

Producing a ton of paper or paperboard from raw timber requires roughly 40 million Btu,[c] the energy equivalent of a ton of crude petroleum. However, more than 40 percent of this energy is derived from wood sources, a proportion that has been increasing over time.

Omitting wood sources of energy, the industries' energy-output ratio declined at a 0.7 percent rate annually between 1947 and 1967. From 1967 to 1974, the ratio declined at an accelerated 1.5 percent rate. However, there was no decline in the ratio between 1974 and 1976. As we shall see, however, the underlying rate of energy conservation has steadily accelerated—and the lack of progress between

---

[a]Both the American Paper Institute and the American Petroleum Institute use the initials API. In this chapter these initials represent the paper trade association, while in Chapter 4 they will represent the petroleum trade association.

[b]Pulp mills that have smaller paper or paperboard mills integrated with them were classified in pulp mills (SIC 2611) prior to 1972.

[c]Energy here is measured in *gross* terms; see Chapter 2 for definition. It should also be noted that all averages are very approximate since energy use varies by process and product, as well as whether it is integrated or nonintegrated.

1974 and 1976 is not necessarily an indication that substantial improvement will not occur in the future.

INDUSTRY CLASSIFICATION PROBLEMS

The definitions used by the Census Bureau to delineate these industries have undergone major changes; in 1954, for example, each pulp mill, including those integrated into a paper or paperboard mill, was counted as a separate plant and classified in the pulp mill industry. In later censuses, a pulp mill and its integrated paper or paperboard mill were classified together as a single plant according to whether shipments of pulp, paper, or paperboard were largest. In 1972, an integrated pulp mill was classified in either the paper or paperboard industry according to whether paper or paperboard was the major product, regardless of the pulp output.

Because of these changes in industry definition as well as variations in the availability of data for these industries, we have used data for the period 1947−67 that cover pulp, paper, and paperboard (SIC 2611, 2621, and 2631) and have treated the three as a single industry.[d] For the period 1967−76, we have omitted independent pulp mills (2611) and treated SIC 2621 and 2631 as a single unit. Since 2611 accounted for only roughly 10 percent of value added and energy use of the three industries combined, we believe that the change in definitions and coverage do not create substantial distortions.

PRODUCTION PROCESSES: OVERVIEW

Although pulp may be made from rags, straw, bagasse, and a variety of other materials, almost all pulp is made from wood or wastepaper, which together account for over 98 percent of all pulp materials.

Most pulp is made from logs. In this case the bark must be removed and the log cut into chips. However, increasing amounts of pulp are being made from waste from lumber mills, which reduces the need for these operations. The chips are then either ground mechanically or mixed with chemicals, cooked, and cleaned. Roughly half of all pulp is then bleached.

If the pulp is to be shipped to another plant for manufacture of paper or paperboard, it must be dried at this point. At an integrated

[d]All three industries (SIC 2611, 2621, 2631) will collectively be called the primary paper industry; the two industries, paper and paperboard (SIC 2621 and 2631), will be called the paper and paperboard industry. It should be kept in mind that the paper and paperboard industry accounts for more than 80 percent of all pulp production, owing to integrated mill production of pulp.

mill, the cellulose and water mixture passes directly through a head box onto a moving screen where water is drained off. This material is then passed through a series of heated rollers to produce paper or paperboard.

MEASURING ENERGY AND OUTPUT

The wood from which wood pulp is made is, of course, itself an energy material, chemically related to sugar (cellulose is a polysaccharide). One could therefore count all wood used in the industry as energy consumption, which would more than double the purchased energy and fuels consumed.

This procedure seems wrong, however, from a number of points of view. First, the energy in wood is not necessarily lost in the manufacturing process; paper and paperboard can be used as energy sources just as wood can be. To attribute energy consumption to the producer rather than the consumer therefore seems inappropriate. Second, because wood no longer plays a significant role as an energy source, conserving wood in the paper industry does not release energy to other industries or sectors.

However, energy from wood is used for heat and power by one manufacturing industry—that is, the primary paper industry. This energy is taken from by-products of the pulp-making process: spent pulping liquors, bark, and waste wood. Spent pulping liquors are particularly significant and represent the industry's largest single source of energy, over 35 percent of all energy consumed for heat and power.

It is very difficult to generalize about energy use in the production process in any industry, and the primary paper industry is no exception. Very roughly speaking, however, pulping operations require between 15 and 25 million Btu per ton, bleaching between 0 and 10 million Btu (0 for unbleached products), and board or papermaking from 10 to 20 million Btu per ton. The gross average is over 40 million Btu per ton, of which about 17 million Btu comes from wood sources.

Moreover, the phrase "Btu per ton of product" has a variety of meanings. For our basic energy statistics, we prefer to use Btu per dollar of value added. This enables us to compare steel production to paper production, for example. But Btu per ton is what is normally used in the literature and in industry discussions. In general, the term refers to a ton of finished (primary) product, including pulping, bleaching, and paper or board production. In 1975, for example, the primary paper industry produced 48 million tons of paper and paperboard, and it consumed some 2.1 quad of energy in captive and purchased energy. This averaged out to nearly 44 million Btu per ton. However, the industry only produced 43 million tons of wood pulp. Several million tons of paper and paperboard were produced from wastepaper or imported pulp. This tended to reduce total energy use because the use of wastepaper or imported pulp requires less energy consumption than production based on pulpwood. Over time, the ratio of domestic pulp production to paper and paperboard production has grown. This has tended to push energy used per ton of paper and paperboard upward.

Having said this much about energy use from wood by-products, we shall say little more. The focus of this book is energy from conventional sources: coal, oil, natural gas, and electricity. Therefore for the rest of this chapter, we shall primarily address purchases of these conventional energy forms. In those cases where we include wood-source energy, we shall be explicit.

HISTORICAL DEVELOPMENTS, 1947–67

From 1947 to 1967, purchased energy use in the pulp, paper, and paperboard industries rose at an annual rate of 3.9 percent. Constant dollar output of the industries rose at an annual rate of 4.7 percent. Thus energy use per constant dollar fell at an average rate of 0.7 percent (Table 3–1). Using tons of pulp, paper, and paperboard as a

Table 3–1. Gross Energy Purchased and Output in Pulp, Paper, and Paperboard (SIC 2611, 2621, and 2631), Selected Years, 1947–67

	1947	*1954*	*1958*	*1962*	*1967*
Gross energy (trillion Btu)	558.7	723.7	800.2	892.5	1206.9
Value added, (billion 1974 dollars)	2373	3165	3674	4630	5926
Ratio gross energy to value added (1000 Btu per 1974 dollar)	235	229	218	193	204
Production of pulp, paper, and paperboard (million tons)	30.55	42.14	49.44	60.54	79.83
Ratio gross energy to tons of product (millions of Btu per ton)	18.3	17.2	16.2	14.7	15.2

Source: *Census of Manufactures*; U.S. Department of Commerce, *Current Industrial Reports.*

measure of production, output rose at an annual rate of 4.9 percent, and energy use per ton fell at an average rate of 0.9 percent.

As we examine the industry more closely, we find that there were exogenous changes that tended to increase the rate of energy use: changes in materials used and products produced. The most important change has been the rapid growth of energy-intensive bleached and semibleached Kraft (sulfate) pulp production (Table 3–2). Almost as important was the relative decline in use of wastepaper as a raw material, a decline that slowed around 1967 (Table 3–3). We estimate that the effects of these two changes have been to increase energy requirements at an average rate between 0.5 percent and 0.6 percent annually. Had it not been for these shifts in products and processes, energy use per constant dollar would have declined from 1947 to 1967 at a 1.2 percent annual rate.

As was mentioned before, the industry derives a substantial amount of its energy use from self-generated fuels, particularly from spent liquors.[e] Growth of this source of energy is closely linked to growth of Kraft pulping, in which a very high proportion of pulping liquors are turned into fuel. For the period before the energy price

Table 3–2. Growth of Kraft Pulping

Year	Total, All Processes	Total Kraft	Bleached Kraft	Semibleached Kraft
	(millions of tons)			
1947	11.9	5.4	0.9	0.2
1954	18.3	9.8	2.7	0.3
1958	21.8	12.3	4.1	0.6
1962	27.9	16.3	5.9	0.9
1967	36.4	23.2	8.7	1.6
1971	43.7	28.8	11.4	1.6
1974	48.3	33.0	13.7	1.7
1975	42.4	29.8	12.9	1.5
1976	48.7	34.5	14.9	1.4
	(percent of total)			
1947	100.0	44.8	7.6	1.6
1954	100.0	53.6	14.7	1.8
1958	100.0	56.5	18.7	2.8
1962	100.0	58.4	21.0	3.3
1967	100.0	63.7	24.1	4.3
1971	100.0	65.9	26.1	3.7
1974	100.0	68.3	28.4	3.5
1975	100.0	70.2	30.4	3.4
1976	100.0	70.7	30.6	3.0

Source: *Current Industrial Reports.*

[e]Spent liquors are the concentrated wastes from the pulping process.

Table 3-3. Decline in Importance of Wastepaper Use in Paper and Paperboard Production

Year	Total Fiber	Wastepaper
	(millions of tons)	
1947	22.8	8.0
1954	28.0	7.9
1958	32.2	8.7
1962	38.6	9.1
1967	47.7	9.9
1971	56.0	11.0
1974	61.3	12.1
1975	53.4	10.3
1976	61.6	12.1
	(percent of total)	
1947	100.0	35.1
1954	100.0	28.0
1958	100.0	27.0
1962	100.0	23.5
1967	100.0	20.7
1971	100.0	19.6
1974	100.0	19.7
1975	100.0	19.4
1976	100.0	19.7

Source: *Current Industrial Reports.*

rises of the early seventies, this was due as much if not more to the value of the chemicals recycled in the process rather than to the fuel value of the spent liquors. Disposal problems were also relevant to attaining the high rate of use of this fuel.

These pulping liquors are less efficient sources of energy than conventional fuels. The measured energy of the liquors, for example, as published in API data, is gross energy. In conversion to steam, some 40 percent of the Btu content is lost even at very modern plants. For the average plant, the proportion lost appears to be much greater.

To illustrate, a modern integrated bleach Kraft mill might require roughly 40 million Btu of steam and electricity (measuring electricity at a gross basis, 10,500 Btu per kwh) per ton of production. Its spent pulping liquor production may be 18 million Btu, but this results in only 10 million Btu of steam. Hence net input requirements are 30 million Btu of gross purchased energy, and total gross energy use (including wood sources) is 48 million Btu.

One means of offsetting Kraft energy costs is to use the spent liquors to generate electricity as well as steam. This technique— known as cogeneration—is extensively practiced in the industry, particularly in modern plants. However, even greater use of this technique is discouraged because sale of electricity is complicated by public regulation of utilities (in most cases, this is because the plant in becoming a seller of electricity would become a "utility producer" and thus fall under public regulation). Partly as a consequence, purchases of electricity increased much more rapidly than self-generation of electricity during this period.

ENERGY CONSERVATION, 1967–76

On the surface, the paper and paperboard industry's recent energy usage appears erratic and difficult to interpret (Table 3–4). For the entire period, the energy-output ratio declined at a 1.1 percent rate, substantially faster than the 0.7 percent rate from the 1947–67 period. Looking at the year-to-year record, however, we see that from 1967 to 1974 the energy-output ratio dropped steadily, while in 1975 it jumped suddenly, and even in 1976 it had not returned to the 1974 level. Thus between 1967 and 1974, the energy-output ratio declined at a 1.5 percent rate, while between 1974 and 1976, it rose at a 0.2 percent annual rate. We shall see that the lack of improvement between 1974 and 1976 is the net result of continued energy conservation by the industry and a combination of factors driving the energy-output ratio upward.

From 1967 to 1976 was a period of ups and downs for the paper and paperboard industry, much as it was for the economy as a whole. In 1967, the industry found itself facing softening demand following the biggest two-year capacity expansion in the post-war period. By 1969, demand had eaten up most of the excess capacity, but the recession of 1970 put the industry in doldrums. With easing of the recession, capacity suddenly became very tight, and the boom years of 1972–73 found the industry scrambling to fill orders. But in 1975, the worst recession since the 1930s sent production nose-diving, idling a fifth of the industry's ability to produce. Although 1976 was a year of recovery, total production was slightly below that of 1974.

We have identified four basic factors in this industry that affect the rate of energy utilization other than the underlying rate of technical progress in energy use and the general price level of energy. These factors are: (1) the rate of capacity utilization; (2) the rate of investment in new plant and equipment; (3) changing process and product mix; and (4) environmental regulation.

Capacity Utilization
The rate of capacity utilization affects energy use per unit of output in a very straightforward way. As capacity utilization falls, energy use falls but less rapidly than production—thus the lower the

Table 3–4. Energy Use and Output, Paper and Paperboard (SIC 2621 and 2631), Selected Years, 1967–76

	Gross Energy (trillion Btu)	Value Added (billion 1974 dollars)	Energy-Output Ratio (1,000 Btu per 1974 dollar)
1967	1,105	5.61	197
1971	1,231	6.66	185
1974	1,294	7.33	177
1975	1,191	6.34	188
1976	1,279	7.16	179

Source: Energy: *Census of Manufactures and Annual Survey of Manufactures;* Value Added: *Census of Manufactures, Annual Survey of Manufactures and Federal Reserve Board, Indexes of Industrial Production.*

rate of capacity utilization, the higher the energy-output ratio. According to the studies of API, a 1 percent decrease in the capacity utilization rate causes an increase in the energy-output ratio of about 0.27 percent.

As can be seen in Table 3—5, capacity utilization was higher in 1974 than in either 1967 or 1976. As a consequence, the energy-output ratio was relatively understated in that year. The second row of Table 3—6 shows the energy-output ratio corrected for capacity utilization. This might be thought of as the technical energy-output ratio of the existing stock of capital equipment in any given year.

Table 3—5. Pulp, Paper, and Board Capacity and Production, Selected Years 1967—76 *(tons of pulp, paper, and board)*

	Capacity	*Change in Capacity*	*Production*	*Capacity Utilization (percent)*
1967	88,953	5,109	79,850	89.8
1971	101,427	2,134	93,988	92.7
1974	111,010	2,040	103,184	93.0
1975	112,284	1,274	90,971	81.0
1976	114,715	2,431	103,428	90.2

Source: American Paper Institute, *Statistics of Paper, 1977.*

Table 3—6. Gross Energy Purchased per Constant Dollar of Output in Paper and Paperboard (SIC 2621 and 2631), Selected Years, 1967—76 *(1,000 Btu per 1974 dollar)*

	1967	*1974*	*1976*
Actual ratio	197	177	179
Ratio, adjusted for capacity utilization	195	177	177
Ratio, plus adjustment for rate of investment	194	178	175
Ratio, plus adjustment for change in mix	194	174	169
Ratio, plus adjustment for pollution abatement	194	173	167

Note: For adjustments, see text, pp. 55ff.
Source: *Census of Manufactures; Annual Survey of Manufactures.*

The Rate of Investment

The rate of investment in new plant and equipment has two successive effects on the energy-output ratio. First, in the year in which a new plant or major plant expansion is made, the energy-output ratio tends to rise because the plant is in a shakedown phase of operation: production is low and energy use is relatively high while the kinks are being worked out. However, in subsequent years the new plant, having been built more efficiently than older plants, reduces the industry energy-output ratio. On the other hand, if the rate of investment falls in a given year, this may temporarily lower the energy-output ratio (because a smaller than normal number of plants are in the shakedown phase). But subsequent years will then show a higher than normal rate of energy use.

The average capacity increase for the period 1966—76 was 3,087 thousand tons. In 1967, the primary paper industry experienced a large increase in total capacity. In 1971 and again from 1974 to 1976, the average capacity increase was less than normal. This put upward pressure on the energy-output ratio in 1967 and had the opposite effect in 1971 and from 1974 to 1975. But by 1976, the effect of previous years outweighed the current effect. In effect, nearly 3 million tons of productive capacity were "missing" compared to the normal run of events—and this was capacity that typically would have been highly energy conserving.

If we assume that these new plants use 30 percent less energy than the average plant in existence, the energy-output ratio was 0.9 percent higher than it would have been had the investment rate been more normal.

Changing Product and Process Mix

The effects of shifts away from wastepaper and toward Kraft (bleached or semibleached) pulp diminished slightly between 1967 and 1974. Recycling of paper and paperboard increased after 1967 as a result of increased scrap collection and greater demand for recycled products, both of which were caused by increased public concern for the environment. Whereas from 1947 to 1967 these factors increased the energy requirements for the industry at an average rate of 0.5 percent to 0.6 percent, their influence diminished to a 0.3 percent rate of increase for 1967 to 1974 and a 0.4 percent annual rate for 1974 to 1976.

Environmental Regulation

Expenditures for environmental protection have had a long history in basic materials processing, and the primary paper industry is no exception. However, expenditures for pollution regulation have taken a dramatic step upward in the past ten years as a consequence of greatly enhanced public concern and new legislation. Since pollution abatement does not increase output (at least, not as output is normally measured), all energy expenditures associated with it increase the energy-output ratio.

Environmental factors added a small amount to 1974 energy use (compared to 1967 practice) and a somewhat larger amount to 1976 energy use (see page 58). The rate of energy savings from 1967 to 1974 is lowered by less than one-tenth of 1 percent, but the rate from 1974 to 1976 is decreased by about 0.3 percent annually.

The fifth row of Table 3–6 shows the rate of energy savings that would have prevailed if these factors had remained constant. For the period 1967–74, the annual rate of decline of the energy-output ratio is 1.6 percent. For 1974 to 1976, the rate of decline is 1.8 percent. As can be seen, the underlying picture is much as one would predict from economic theory—as energy prices rise, energy savings increase. However, for the reasons we have been discussing, these energy savings were not realized; barriers to energy conservation proved greater than incentives for the period 1974–76.

AMERICAN PAPER INSTITUTE ENERGY STATISTICS

Additional information on energy use is available from the surveys of the API. Monthly data exist for 1972 and from the fall of 1974 onward. Data were also collected for 1971 but unfortunately are not comparable. These surveys are less complete than the data collected by the census in the proportion of the industry responding. However, they include a broader range of information and are the only source of figures for self-generated fuels. The data sets for 1972, 1975, and 1976 are each somewhat different, reflecting different response rates and purposes. Nevertheless, they agree substantially with the trends appearing in Census Bureau figures.

Table 3–7 shows the API data as reported to the federal government under the Voluntary Industrial Energy Conservation Program (VIECP). In this table, electric energy has been converted to a gross basis.

Detailed agreement on fuels between the two surveys is highly unlikely because coverage varies. However, it should be noted that the API survey shows much greater purchases of residual fuel oil than does the *Annual Survey of Manufactures*. Because the *Annual Survey* has had difficulty with fuel oil in the past (the 1974 fuel oil data

Table 3–7.   Detailed Comparison of American Paper Institute and Census Bureau Energy Data, Selected Years, 1972–76

(trillion Btu)

	American Paper Institute			Census Bureau 1976
	1972	1975	1976	
Purchased electricity	284.5	279.8	336.6	321.9
Purchased steam	21.8	16.1	18.0	
Coal	220.0	174.0	212.3	203.4
Residual fuel oil	466.5	437.9	483.0	358.0
Distillate fuel oil	20.3	13.5	17.7	58.2
Liquid propane gas	2.3	1.3	1.7	
Natural gas	461.9	347.6	327.6	291.8
Other	.2	1.0	2.5	46.1
Energy sold	(–28.3)	(–34.3)	(–37.4)	
Total purchased fossil fuel and energy	1451.2	1236.9	1372.0	1279.4
Hogged fuel	44.8	56.1	75.6	
Bark	102.6	82.8	95.0	
Spent liquor	748.4	675.8	754.3	
Self-generated hydroelectricity	27.0	27.9	29.3	
Other self-generated fuel	3.4	6.1	6.2	
Total self-generated and waste fuels	926.2	848.6	960.5	
Total energy	2377.4	2085.5	2332.5	

Note: API data include SIC 2611, whereas Census Bureau data exclude it. API data are estimated rates based on samples representing 81 percent of the industry's output in 1972, 80 percent in 1975, and 85 percent in 1976. Energy is measured in gross energy terms. Census Bureau energy data for 2611 have not been published as of this writing.
Source: American Paper Institute; *Annual Survey of Manufactures, 1976.*

were later extensively revised), it would appear that there are still difficulties in collection of these figures.

Energy totals cannot be compared directly because API data include, while the census data exclude, nonintegrated pulp mills. The more relevant comparison is the trend of total purchased fossil fuel and electric energy. To facilitate this comparison, we have estimated in Table 3—8 1972 energy use based on Census Bureau data (see footnote to Table 3—8).

The rate of decline of total purchased energy for the API surveys was 1.4 percent annually from 1972 to 1976, compared with the Census Bureau measure of 1.6 percent. The 1972 to 1975 rates of decline diverge more sharply (4.4 percent for the Census Bureau, 5.2 percent for API). Some of this divergence may be due to the fact that API has lower coverage (80 percent to 85 percent) that is nonrandom. In unusual conditions (such as the 1975 recession) nonrandom samples are generally less reliable; because they are not random, they will often move differently from the rest of the sample universe.

Table 3—8. Comparison of Trends in American Paper Institute and Census Energy Data, Selected Years, 1972—76

	1972	*1975*	*1976*
API			
Gross energy (trillion Btu)	1451	1237	1372
Output (million tons)	66.9	58.4	66.9
Energy-output ratio (million Btu per ton)	21.7	21.2	20.5
Census			
Gross energy (trillion Btu)	1363	1191	1279
Output (billion 1974 dollars)	7.06	6.34	7.16
Energy-output ratio (1000 Btu per 1974 dollar)	193	188	179
Indexes (1972 = 100)			
API gross energy	100.0	85.2	94.5
Census gross energy	100.0	87.4	93.8
API energy-output ratio	100.0	97.6	94.6
Census energy-output ratio	100.0	97.4	92.7

Note: 1972 energy data are published in *Annual Survey of Manufactures, 1974.* This source provides kwh of purchased electricity and total purchases of fuels in dollars, but not physical units of fuels. Total fuel Btu for 1972 were estimated by updating the 1971 ratio of fuel Btu per dollar using the BLS Wholesale Price Index for fuels. This provided a 1972 estimate of Btu per dollar for fuels, which multiplied times the 1972 purchases yielded an estimate of total Btu.

Source: Table 3-5; Table 3-7; American Paper Institute; *Annual Survey of Manufactures.*

While these comments indicate that the API data may be superior in some respects to the data collected by the Census Bureau, we have based our interpretations primarily on the latter because they go back to 1947 and because, despite all the difficulties, they agree in trend with the API data for the period 1972–76.

One further reason for using the census data is that API defines output as tons of machine-dried pulp, paper, and paperboard, that is, paper and paperboard output plus pulp that is dried for shipment. Thus, the API definition counts two tons for every ton of noninte-grated production of pulp and paper or pulp and board and one ton for every ton of integrated production. The API survey data would be of more general value if greater detail were published on output.

ENERGY USE FOR ENVIRONMENTAL PROTECTION

A 1975 survey by the National Council of the Paper Industry on Air and Stream Improvement (NCASI) provides the basis for our esti-mates of environmental energy costs in the paper industry. This sur-vey measures energy costs due to environmental protection since 1960 on a useful energy basis (electricity measured at 3,413 Btu per kwh). Converting the data to a gross energy basis, we estimate that additional energy use between 1974 and 1976 was roughly 10 trillion Btu. This is 0.6 percent of total annual purchased energy for the in-dustry.

CONCLUSIONS

The paper and paperboard industries, treated in this chapter as a single industry, markedly accelerated energy savings after 1967. However, in the period after the Arab oil embargo, 1974 to 1976, the energy-output ratio rose.

One major cause of this change in trend is the 1975 recession, which left production in 1976 below that of 1974. The general slow-down in economic activity, by lowering demand for the paper and paperboard industry's product, has reduced the rate of capacity utili-zation and also the rate of new investment. Both of these factors have pushed the energy-output ratio upward. As the recovery contin-ues, these factors should have diminishing weight. Other factors that caused the energy-output ratio to rise include environmental regu-lation and the continuing shift toward Kraft production. In the absence of these factors, it appears that there would have been a modest acceleration in the rate of decline of the energy-output ratio after 1974.

✳ *Chapter 4*

Chemicals and Petroleum Refining

Organic chemicals, inorganic chemicals, and petroleum refining are combined in this chapter because each uses energy primarily to bring about chemical reactions. In addition, substantial amounts of energy are used in each industry for distillation (the separation of chemical streams into salable products).

In 1974, these three industries together accounted for 3.7 quad (quadrillion Btu) of purchased energy (Table 2–2), which is roughly 21 percent of purchased energy used by all manufacturing industries. As is pointed out in the following sections, these industries also use nonpurchased forms of energy for heat and power and fuels for raw materials. As a consequence, the total absorption of energy by these industries is roughly 9 quad, more than one-third of all energy used in manufacturing.

INDUSTRIAL ORGANIC CHEMICALS, N.E.C. (SIC 2818)

In 1974, the industrial organic chemical industry used 1.2 quad of energy for heat and power. Because this industry is based on petrochemical feedstocks, it also uses energy as a raw material. It is estimated that between 1.2 and 1.5 quad of energy was used in 1974 for feedstocks. Whether we measure the industry's total energy use as purchased fuels and electric energy or as total energy absorption, it is the third largest industrial energy user, ranking behind blast furnaces and steel mills and petroleum refining.

Background

The industrial organic chemical industry produces petrochemicals, chemicals derived from petroleum and natural gas. Industry products do not, however, include cyclic intermediates (major petrochemicals based on benzene, toluene, and xylene) and basic petrochemicals produced in petroleum refineries and natural gas plants. The chemicals produced in the industry typically involve an elaboration of chemicals—simple petrochemicals like ethylene, ethane, and propylene are combined with other simple chemicals such as oxygen, nitrogen, and chlorine, and the resulting products are further combined. Most of the products of the industry are intermediate commodities, used to make synthetic fibers and plastics, for example, instead of being directly purchased by consumers.

The industry has been characterized by an enormous growth of processes and products. While the consumer encounters these petrochemical advances primarily in an increasing array of synthetic materials, the industry itself undergoes a constant further process of change by finding new chemical routes to existing final products. For example, acetic acid can be produced from acetaldehyde, itself a product of ethyl alcohol, in turn a product of ethylene, or it may be produced directly from butane, a natural gas liquid. Now it is increasingly produced from methanol, a derivative of natural gas.

Organic chemical processes tend to require substantial amounts of energy, first in elevation of temperature and pressure to promote the speed of the reactions, and second in distillation, separating the finished product from by-products and from unreacted raw materials.

Each new chemical and process tends to undergo a long history of industrial development, beginning from an early stage of relatively inefficient production methods, usually batch processing, and eventually reaching highly efficient large-scale continuous production if it finds or makes a large enough market.

This dynamism has enabled the industry to save energy at a rapid pace. Cost savings of all kinds have in fact been quite general in the industry—technological advance has generally kept industry price increase well below the average for the economy, and this steady relative price reduction has enabled the industry to penetrate a wide variety of markets and maintain a rate of growth far above the national average.

However, new chemicals and processes are initially inefficient and hence relatively energy intensive. Their continual introduction thus slows the rate of decline of the total energy-output ratio. For example, from 1954 to 1967 this ratio dropped at a 0.7 percent annual rate. From 1967 to 1974, the rate accelerated to 5.1 percent. However, between 1974 and 1976, because of declining capacity utilization, the energy-output ratio rose.

These data include energy costs of new chemicals and processes. The industry has undoubtedly saved energy more rapidly when measured for each chemical. By using this basis, it is possible that energy savings from 1954 to 1967 were close to 3 percent rather than less than 1 percent. We have not in this book been able to measure energy savings on this basis precisely. However, it seems likely that energy savings chemical by chemical have accelerated substantially since 1967.

Measurement of Output and Energy

As noted earlier, energy is used in this industry both for heat and power and for raw material feedstocks. As in other sections in this book, we concentrate on energy used for heat and power rather than the raw material use of energy. Unfortunately (from an analytical standpoint), energy derived from feedstocks is a frequent source of heat or power in the industry. This takes two main forms. One is that the chemical reaction that produces a particular chemical may be exothermic, that is, heat generating. All "partial oxidation" processes, for example, simply mean partial burning of the feedstocks. Thus, the energy needed to sustain a chemical reaction may come from the reaction itself. Moreover, there may be surplus heat, which can then be used in other processes.

The other way in which feedstocks can be used for heat or power is in the generation of by-products that may be used as fuels. The production of ethylene from ethane produces by-product hydrogen and methane, which may then be burned.

The impcrtance of these uses of feedstocks can be seen by noting that ethylene in the above example uses 13.4 thousand Btu per pound, of which 11.6 thousand Btu is obtained from feedstocks. As a consequence, ethylene production absorbs the equivalent of one-fourth of all purchased energy used for heat and power in the industry, but it obtains the energy almost entirely from its feedstock. Ethylene therefore has very little net effect on the industry's purchases of energy for heat and power despite the fact that it is produced in huge volumes.

Output measures are also a source of difficulty in the chemical industry. One measure frequently used is total pounds produced. The problem here is that process variation can completely eliminate the production of intermediate chemicals. To take a simple example, acetaldehyde was once made exclusively from ethyl alcohol, which is in turn made from ethylene. Now acetaldehyde is primarily made directly from ethylene. The elimination of the intermediate chemical reduces the total pounds produced, but the beginning and end product remain the same. The same final output is achieved, but with less measured production.

The method used to calculate the census output index,[a] which largely employs data on sales, appears to compound the problem because movements in sales can reflect changes in degree of integration of the industry as well as changes in process. In the example above, an ethylene producer can start an acetaldehyde plant, using the ethylene to make the acetaldehyde and no longer selling it. The ethylene then disappears from sales data.

There appears to be no fully satisfactory way out of this dilemma. It should be recognized, however, that as a consequence it is probable that both the output measures being used—pounds of production data from the International Trade Commission and our Census Bureau-Federal Reserve Board (FRB) production index—understate true output.[b]

Two further data anomalies must be mentioned. Chlorine is a product classified in SIC 2812, alkalies and chlorine. Ethylene and propylene are classified as products of petroleum refineries, SIC 2911. But when any of these three chemicals is produced in integrated organic chemicals plants, census energy data include energy used to produce them. Some 60 percent of all chlorine and 40 percent of all ethylene is produced in plants classified in SIC 2818. The substantial energy use represented by these chemicals must be allowed for in our analysis.

Historical Developments, 1954–67

From 1954 to 1967, the organic chemicals industry grew at a 10.0 percent annual rate. At the same time, energy use grew slightly slower, that is, at 9.2 percent annually. Consequently, the energy-output ratio fell at a steady 0.7 percent average annual rate (Table 4–1).

If the influence of chlorine and ethylene is removed, it can be seen that energy per pound of production drops at a 0.8 percent rate (Table 4–2). This slight acceleration occurs because captive chlorine production, which is highly energy intensive, grew more rapidly than did production generally.

[a]See Appendix B for further information on Census Bureau–FRB output indexes.

[b]The 1967-72 index number for organic chemicals had unusual problems that have been adjusted for in this chapter but not in Chapter 2. See Appendix B for a fuller discussion.

Table 4–1. Energy Use and Output in Industrial Organic Chemicals, n.e.c., Selected Years, 1954–67

	Gross Energy (trillion Btu)	Value Added (billion 1974 dollars)	Energy-Output Ratio (1,000 Btu per 1974 dollar)
1954	313	1.23	255
1958	418	1.71	245
1962	642	2.71	236
1967	982	4.22	233

Source: *Census of Manufactures.*

A group of twenty petrochemicals were studied in detail (see Appendix C). These large-volume petrochemicals account for some 60 percent, by weight, of all chemicals produced in this industry. Detailed estimates of energy use by each chemical were prepared for 1954, 1967, and 1980. As can be seen in Table A–2, these chemicals accounted for only about one-fifth of total energy use. Yet between 1954 and 1967, these large-volume petrochemicals accounted for all of the decline in energy use. Their energy use per pound declined at a 3.0 percent annual rate, and for the remaining chemicals energy use per unit actually increased slightly.

The reason why the residual group shows little energy savings lies in the dynamics of the industry. A major cause of the rapid growth of the petrochemical industry is a research effort that has generated a continuous stream of new chemical products. These new products, initially produced in small volumes by relatively inefficient methods, represent a continuing influx of energy-intensive chemicals.

If we could remove the newly introduced chemicals from the older, small value chemicals in the residual group, it seems likely that the remaining chemicals would themselves show a high rate of annual energy savings, much like their large-volume relatives. Unfortunately, this cannot be done.

It seems possible then that the average rate of energy savings for each chemical in the industry was close to 3.0 percent. If we also recall that our output measure is believed to be understated, the rate of energy savings is yet higher.

Energy Conservation, 1967–76

Output of the organic chemical industry rose from 1967 to 1974 at the lively rate of 9.3 percent annually as measured by the combined Census Bureau–FRB index. It should be mentioned that while our output index grew at the same rate as total pounds of production from 1954 to 1967, this did not hold true for the later period. Total

Table 4–2. Detailed Energy Use and Output, Industrial Organic Chemicals, n.e.c., 1954 and 1967

	Gross Energy (trillion Btu)		Production (billion lbs.)		Energy-Output Ratio Production Basis (1,000 Btu per lb.)		Annual Rate of Change, Energy-Output Ratio
	1954	1967	1954	1967	1954	1967	1954–1967
Total	313	982					
Chlorine	40	165	7.4	7.2	28.6	22.9	-1.7%
Ethylene	10	10	2.0	5.6	5.0	1.8	-7.5%
	263	807	18.3	62.4	14.3	12.9	-0.8%
Twenty petrochemicals	76	178	10.4	36.3	7.3	4.9	-3.0%
Other chemicals	187	629	7.9	26.1	23.7	24.1	+0.2%

Source: *Census of Manufactures*; U.S. International Trade Commission, *Synthetic Organic Chemicals*; J.G. Myers, *Energy Consumption in Manufacturing.*

pounds of production grew by 7.6 percent—still rapid but well below the rate of our index (Table 4−3). The principal reason for this difference appears to be that the plants in this industry produced increasing amounts of secondary products (products that are not classified in the industry definition and are produced mainly by plants in other industries).[c]

An analytic problem is created because of the divergence between industry output and total pounds produced of the industry's primary products. Whereas between 1954 and 1967 we were able to relate energy consumption directly to primary product production, an adjustment for the post–1967 period had to be made. We have assumed that the secondary products have the same energy intensity as the primary products, and whenever we compare energy use to production in pounds, we reduce the total energy use by the ratio of total output to primary products (the ratios shown in column 4 of Table 4−3).

Between 1967 and 1974, energy use[d] per constant dollar dropped a total of 31 percent, an annual rate greater than 5 percent. 1974 was an exceptionally good year, one in which plants operated at greater than normal rates of production. At the same time, because of the embargo-induced rationing of petroleum products and the natural gas shortage, a number of companies took unusually vigorous steps to reduce plant energy use. Energy price rises were an additional impetus to conservation.

Then the 1975 recession struck. Plants that had been operating at full tilt in 1974 reduced their utilization rates to as low as 70 percent of capacity. As an illustration, we note that a fire early in 1974 at a Texas petrochemical complex put a major unit out of operation for three months, but the plant was able to maintain its output since capacity had been previously only three-fourths utilized. One factor that kept capacity utilization rates low was continued expansion by the industry. Many observers had expected the recession to be quite mild—the *Oil and Gas Journal* in March expected production in late 1975 to exceed 1974 levels—and this factor kept construction going through most of 1975.

The importance of utilization rates is that plants require more energy for each unit of production when operated at less than full capacity. Unfortunately, measurement of capacity utilization in the

[c]The census output index, like census energy data, is based on the plants classified in the industry; pounds of production is a measure of output of products classified in the industry.

[d]Energy data for the 1974–76 period include estimates for urea. Since 1972, urea has been classified in SIC 2873, nitrogenous fertilizers.

Table 4–3. Output, Industrial Organic Chemicals, n.e.c., Selected Years, 1967–76

	Production (million lbs.)	Production Index	Census Bureau-FRB Output Index	Ratio Output Index Production Index
1967	62,385	100.0	100.0	1.000
1971	82,509	132.3	145.3	1.098
1974	104,432	167.4	186.8	1.116
1975	89,679	143.8	163.1	1.134
1976	102,868	164.9	192.7	1.168

Note: The Census Bureau-FRB index has been adjusted for changes in the ratio of sales to output, see Appendix B for further details.

Source: *Synthetic Organic Chemicals; Indexes of Industrial Production; Census of Manufactures, 1972.*

industry as a whole is very difficult—the industry is too diverse for a single measure to be very meaningful. For this reason, we have been unable to establish a direct quantitative relationship between capacity utilization and energy savings.

Output improved considerably in 1976, but it was still not as high (in terms of pounds of primary production) as it had been in 1974. Because of the continuing expansion of capacity, utilization in 1976 was clearly well below that of 1974.

As can be seen in Table 4−4, the energy-output ratio rose 6.8 percent in 1975 as a direct result of the recession. Although the ratio resumed its decline in 1976, it did not reach the low level of 1974.

Table 4−5 shows a breakdown of energy use, in which the influence of chlorine and ethylene production is first removed, and then the results of our detailed study of twenty major petrochemicals are presented. Estimates for 1974 to 1976 for the twenty petrochemicals are based on the assumption that production is at normal capacity levels. Because of the heavy influence of the recession in 1975, the results for that year are unreliable and should be seen as very gross estimates. Although the data for 1976 also involve much estimation, they are on firmer ground.

It is interesting to note that the twenty petrochemicals show very little change in energy use per unit between 1974 and 1976 (Table 4−6). This arises for a combination of reasons, among them shifts to somewhat more energy-intensive processes (like the Oxirane process for producing propylene oxide) for a few chemicals.

Of even greater interest is that for the entire period 1967−76, the smaller volume petrochemicals have sharply accelerated their energy savings. Whereas in the 1954−67 period the smaller volume petrochemicals had shown no gains, they now show very sharp gains. While it must be admitted that the estimates contain errors, the strength of the result can be shown by the following calculation. If we assume that our 1976 energy-output ratio is too high by 20 percent for the large-volume petrochemicals so that the energy-output ratio for this group declined at a 5 percent rate instead of 3 percent, the rate of change for the residual group is still nearly 3 percent.

New processes and products continued to be introduced during this period, but it is possible that the rate of such introduction as well as its relative importance in energy use is declining. Since data on energy use by new chemicals are not obtainable, it is difficult to determine how much of the improvement to attribute to increased energy savings for each chemical and how much to a slower rate of innovation (or lower energy costs for innovation.)

Table 4–4. Energy Use and Output in Industrial Organic Chemicals, n.e.c., Selected Years, 1967–76

	Gross Energy (trillion Btu)	Value Added (billion 1974 dollars)	Energy-Output Ratio FRB Basis (1,000 Btu per 1974 dollar)	Annual Rate of Change Energy-Output Ratio (change from previous period)
1967	982	4.22	232.7	
1971	1,095	6.13	178.6	- 6.4%
1974	1,270	7.88	161.2	- 3.4%
1975	1,185	6.88	172.2	+ 6.8%
1976	1,355	8.13	166.7	- 3.2%

Source: *Census of Manufactures; Annual Survey of Manufactures;* Table 4–3.

Table 4–5. Detailed Energy Use and Output in Industrial Organic Chemicals, n.e.c., Selected Years, 1967–76
(gross energy in trillion Btu; production in billion pounds)

	Total (including ethylene and chlorine)	Chlorine	Ethylene	Net (excluding ethylene and chlorine)	Twenty Petrochemicals	Other Chemicals
1967 Gross energy	982	165	10	807	178	629
Production		7.2	5.6	62.4	36.3	26.1
1974 Gross energy	1,138	306	10	822	190	593
Production		14.6	10.0	104.4	61.4	43.1
1975 Gross energy	1,045	268	10	767	181	586
Production		12.2	8.6	89.7	54.2	35.5
1976 Gross energy	1,159	290	10	859	218	641
Production		13.8	9.8	102.9	59.5	43.4

Source: Table 4–4; *Synthetic Organic Chemicals*; J.G. Myers, *Energy Consumption in Manufacturing.*

Table 4–6.   Detailed Energy-Output Ratios, Industrial Organic Chemicals, n.e.c., Selected Years, 1967–76
(1,000 Btu per pound)

	Chlorine	Ethylene	Net (excluding ethylene and chlorine)	Twenty Petrochemicals	Other Chemicals
1967 Energy-output ratio	22.9	1.8	12.9	4.9	20.5
1974 Energy-output ratio	21.0	1.0	7.9	3.7	13.8
1976 Energy-output ratio	21.0	1.0	8.3	3.7	14.8
Rates of change 1967 to 1976	-1.0%	-6.3%	-4.8%	-3.1%	-3.6%

Source: Table 4–5.

Manufacturing Chemists Association
Energy Data

There are no trade association reports on energy use that cover organic chemicals as a four-digit industry. The same holds true for the industry in the following section, inorganic chemicals. The closest approximation is the Manufacturing Chemists Association (MCA), which reports data on all of SIC 28, chemicals and related products.

SIC 28, in addition to including both organic and inorganic chemicals, includes plastics, synthetic fibers, drugs, fertilizers, synthetic rubber, paints, and soaps. Because of the wide range of the industry at the two-digit level and because the companies reporting to the MCA make products in all of these product groups, it is not possible to compare the data presented in MCA reports with the data presented here. It should be noted that the MCA reports agree in trend with data for all SIC 28. Between 1974 and 1976, the MCA shows roughly a 2 percent annual rate of decline for the energy-output ratio, the same as our rate for SIC 28.

Energy Costs for Pollution Abatement

Direct estimates of pollution abatement energy costs for the organic chemical four-digit industry do not appear to exist. However, the MCA reports do include estimates for all of SIC 28. These estimates indicate that both environmental regulation and occupational safety and health requirements increased energy use at about a 0.3 percent rate annually between 1974 and 1976.

Conclusions

A detailed study of twenty petrochemicals, accounting for 60 percent of all pounds produced, is included as part of this book. In the 1954–67 period, these chemicals accounted for all energy savings. Energy savings for the residual group were offset by introduction of new chemicals and processes.

Since 1967, savings appear to be quite general. Part of this result may be due to sharply lowered energy use in introducing new processes and products. Energy savings overall accelerated from 0.7 percent during the 1954–67 period to 5.1 percent during the 1967–74 period. But primarily as a result of decreases in rates of capacity utilization, the energy-output ratio rose from 1974 to 1976.

INDUSTRIAL INORGANIC CHEMICALS, N.E.C. (SIC 2819)

The industrial inorganic chemical, n.e.c., industry produces a wide range of chemicals. It is a catchall industry, defined in large part by what it does not include. Its definition excludes organic chemicals (chemicals containing carbon and hydrogen), alkalies and chlorine, elemental gases, and dyes and pigments.

Included in the statistics for the inorganic chemical industry are a number of government owned but privately operated plants. Three of these, the uranium diffusion plants run by the Department of Energy (DOE), are of special interest from the perspective of energy use. In 1975, these three plants, whose function is to enrich milled uranium to levels required for power plant use or weapons use, consumed over one-third of all energy used by the industry. Since this use is so important and at the same time unique, we shall present separate figures for it and generally exclude it from the discussion of the industry at large. Unless otherwise specified, "inorganic chemical industry" means the industrial inorganic chemical industry, n.e.c., excluding uranium diffusion plants.

The inorganic chemical industry (excluding diffusion plants) used roughly 600 trillion Btu of energy for heat and power in 1974. In addition, the industry—which has historically produced roughly three-fourths of all U.S. ammonia—consumed between 275 and 350 trillion Btu of natural gas for feedstock in the production of ammonia. (Ammonia does not contain carbon, and thus it is not an organic chemical even though its feedstock is natural gas.)

Between 1967 and 1974, output of the uranium diffusion plants rose at a slow 1.3 percent rate. However, between 1974 and 1976, the plants' output rose at a 21.5 percent clip. In addition, the energy-output ratio increased at a 2.3 percent rate. As a consequence, the industrial inorganic chemical, n.e.c., industry (including the diffusion plants) energy-output ratio rose between 1974 and 1976. However, if we exclude the diffusion plants, a very different story emerges.

Between 1967 and 1974, the industry clearly accelerated an already substantial rate of energy saving and attained an even higher rate from 1974 to 1976. The annual rate of energy savings for the industry less diffusion plants was 1.9 percent from 1954 to 1967. From 1967 to 1974, the rate accelerated to 2.8 percent and from 1974 to 1976 even further to 3.8 percent.

Uranium Diffusion

Natural uranium contains only 0.7 percent of the radioactive isotope U−235. In order to fuel a nuclear power plant, the uranium must have a U−235 content of at least 2 percent. Attaining this enrichment is the task of the three uranium diffusion plants that are included in this industry. In 1974, these three plants used about 0.3 quad of energy, nearly all of it electrical.

A major project—the cascades improvement program and the capacity uprating program—is now underway to increase capacity and decrease the rate of energy use. This is the first major change in the plants since they were built in the 1940s and early 1950s.

During the 1950s production of the diffusion plants was primarily for nuclear weaponry. As a consequence, production data were classified as confidential and excluded from census output indexes. To obtain an output index for the industry including diffusion plants, it is necessary to use diffusion plant energy data to estimate diffusion plant production. In the recent period (Table 4—7), the energy-output ratio of diffusion plants has increased at a 0.6 percent annual rate. However, we believe that this increase is a relatively new phenomenon and that the historical energy-output ratio was roughly constant. The estimates in Table 4—8 are based on the assumption that the energy-output ratio from 1954 to 1967 was constant.

It should be mentioned that alternative means of producing enriched uranium currently under development promise drastic reductions in energy use. Centrifugal, nozzle, and laser enrichment techniques appear to use between a tenth and a twentieth as much energy as does the diffusion technique.

Energy and Output Measurement

In 1972, the definition of this industry was changed by the Office of Management and Budget: ammonia, nitric acid, and ammonia compounds were classified in nitrogenous fertilizers (SIC 2873), phosphoric acid moved to phosphatic fertilizers (SIC 2874), and household bleaches moved to polishes and sanitation goods (SIC 2842). From the perspective of measuring energy use, the only major energy user that is reclassified is ammonia. Fortunately, almost all energy use in the new industry SIC 2873 is due to either ammonia or urea production, and less than 10 percent is urea energy consump-

Table 4—7. Energy Use and Value Added in Uranium Diffusion Plants, Selected Years, 1967—76

	1967	*1971*	*1974*	*1975*	*1976*
Energy (trillion Btu)	262.1	206.6	288.3	372.7	446.3
Value added (billion 1974 dollars)	0.686	0.532	0.752	0.956	1.110
Energy-output ratio (1,000 Btu per 1974 dollar)	382	388	384	390	402

Source: Energy Research and Development Administration.

Table 4–8.   Energy Use and Output in Industrial Inorganic Chemicals, n.e.c., Selected Years, 1954–67

Coverage	1954	1958	1962	1967
Industrial Inorganic Chemicals, n.e.c. (SIC 2819), entire industry				
Gross energy (trillion Btu)	695	1050	993	932
Value added (billion 1974 dollars)	1.93	3.11	3.40	3.43
Energy-output ratio (1,000 Btu per 1974 dollar)	360	338	292	292
Industrial Inorganic Chemicals, n.e.c., less diffusion plants				
Gross energy (trillion Btu)	400	494	511	670
Value added (billion 1974 dollars)	1.27	1.73	2.15	2.74
Energy-output ratio (1,000 Btu per 1974 dollar)	315	286	238	245

Source: *Census of Manufactures*; Energy Research and Development Administration.

tion. In addition, relatively reliable information on energy consumption exists for both processes and fits the actual census energy data quite well. For this reason, we decided to estimate the portion of energy shown in SIC 2873 and 2874 that was consumed in production formerly classified in industrial inorganic chemicals, n.e.c. Adding this and corresponding estimates of value added back into SIC 2819 for 1974, 1975, and 1976 allows us to construct a consistent historical series for the industry. In this case, estimates for energy use are more accurate than those for value added. However, it seems unlikely that the errors in the estimates affect trends by as much as 0.1 percent annually.

Historical Developments, 1954–67

Between 1954 and 1967, the inorganic chemical industry reduced its energy use per unit of output at a 1.9 percent annual rate (Table 4–8). Actual energy use increased at a 4.0 percent annual rate, but output grew much faster at a 6.1 percent annual rate.

Two chemicals dominate energy use in the industry: ammonia and aluminum oxide (alumina). Together, these two chemicals accounted for an estimated 38 percent of industry energy use in 1967, 22 percent by ammonia and 16 percent by alumina.

In the making of alumina, there are two major steps. First, aluminum hydroxide is produced from bauxite by the Bayer process. Then aluminum oxide is made from aluminum hydroxide by calcination (roasting in a rotary kiln to drive off water.) Because the calcination step accounts for the bulk of energy used in alumina production, we are primarily concerned with calcined alumina in this book. Almost all calcined alumina (96 percent in 1974) is used to produce aluminum.

Alumina production has generally grown at about the same rate as the inorganic chemical industry, a 5.9 percent rate between 1954 and 1967 (Table 4–9). This pattern, as we shall see, does not hold true

Table 4–9. Output of Ammonia and Alumina, 1954 and 1967
(1,000 short tons)

	Ammonia[a]	Alumina
1954	2736	2886
1967	12194	6046
Annual growth rate	12.2%	5.9%

[a]Substantial amounts of ammonia are produced by plants outside SIC 2819.
Source: *Current Industrial Reports.*

for the period after 1967 because of a sharp drop in aluminum production in the 1975 recession and increasing alumina production abroad.

Ammonia is used directly as a fertilizer and as an intermediate in the production of nitrogen fertilizers such as urea. It is also used in the production of explosives and is a familiar ingredient in household cleaners. Throughout this period, ammonia production grew more rapidly than the industry as a whole. This growth was made possible by large increases in worldwide demand for fertilizer materials, by technological advances in ammonia production that lowered prices, and by stable input prices for ammonia's major raw material and fuel, natural gas. As a consequence, ammonia production rose 12 percent annually, more than quadrupling in volume in thirteen years.

Because the production of ammonia is more energy intensive (by a factor of 5) than the industry excluding ammonia and alumina, ammonia's rapid growth has tended to force the energy-output ratio upward, all other things held equal. Had there been no technical progress, the industry energy-output ratio would have increased at an average rate of 0.4 percent annually because of the shift in output mix.

Energy Conservation, 1967−76

As can be seen in Table 4−10, the energy-output ratio for the industry including diffusion plants declined between 1967 and 1974, then increased between 1974 and 1975, and in 1976 the energy-output ratio was still above that in 1974. This is due to the very rapid growth of energy-intensive uranium enrichment between 1974 and 1976.

Eliminating the diffusion plants, we see that the rate of energy savings for the rest of the inorganic chemical industry accelerated quite strongly. From 1967, the index declined at a 2.8 percent annual rate and from 1974 to 1976 at a 3.8 percent annual rate. Overall, these strong savings in energy use were achieved despite the fact that this was a period of relatively slow growth for the industry. From 1967 to 1976, the industry grew only 2.4 percent annually while total energy use fell at a 0.7 percent rate.

One reason for the strong performance of the industry is the slower growth rates of ammonia and alumina. Ammonia continued to outpace the industry average, but just barely; its growth rate in this period is 3.8 percent annually (Table 4−11). This is only 1.4 percentage points above the industry rate, compared to the 1954−67 period, when ammonia's growth exceeded the industry's by 6.1 percentage points. The alumina growth rate fell steadily behind the in-

Table 4–10. Energy Use and Output in Industrial Inorganic Chemicals, n.e.c., Selected Years, 1967–76

Coverage	1967	1974	1975	1976
Industrial inorganic chemicals, *n.e.c. (SIC 2819), entire industry*				
Gross energy (trillion Btu)	932	965	1035	1080
Value added (billion 1974 dollars)	3.43	4.11	4.12	4.51
Energy-output Ratio (1,000 Btu per 1974 dollar)	271.8	234.8	251.2	239.4
Industrial inorganic chemicals, n.e.c., *less diffusion plants*				
Gross energy (trillion Btu)	670	677	666	634
Value added (billion 1974 dollars)	2.74	3.36	3.16	3.40
Energy-output ratio (1,000 Btu per 1974 dollar)	245	201	211	186

Source: *Census of Manufactures;* Energy Research and Development Administration; *Indexes of Industrial Production.*

Table 4—11.   Value Added by Product, Industrial Inorganic Chemicals, n.e.c., Excluding Diffusion Plants, Selected Years, 1967—76

	Alumina	Ammonia	Other Inorganic Chemicals	Total
	(billion 1974 dollars)			
1967	0.25	0.30	2.19	2.74
1974	0.29	0.40	2.66	3.36
1976	0.25	0.42	2.73	3.40
	(percent of total)			
1967	9	11	80	100
1974	9	12	79	100
1976	7	12	81	100

Source: *Current Industrial Reports; Indexes of Industrial Production;* Aluminum Association.

dustry, and in 1976—in the wake of the 1975 recession—alumina production was equal to its 1967 output.

Averaged together, the two chemicals grew at about the rate of growth of the industry—so instead of being a drag on the energy-output ratio, output mix has virtually no net effect considering this period as a whole.

However, alumina production fell rapidly between 1974 and 1976. For this reason, there were different effects on the subperiods 1967—74 and 1974—76. In the first, the effect of the change in output mix was to raise energy use by about 0.1 percent annually. In the 1974—76 period, however, the decline in alumina as a proportion of the industry reduced the rate of energy use by 0.5 percent annually. Without these effects, the industry savings rate would have been 2.9 percent in 1967—74 and 3.3 percent in 1974—76.

We are fortunate in having fairly detailed data available for 1974 and 1976 for both alumina and ammonia. The data for alumina are from the Aluminum Association; the data for ammonia are estimated from Census Bureau energy data and Fertilizer Institute surveys. As can be seen, the energy-output ratios for both alumina and ammonia *rise* between 1974 and 1976 (Table 4—12).

The increase in the energy-output ratio for alumina appears to be due to the low rate of capacity utilization in the industry. Although part of the reason for this low rate of production was the recession of 1975, increasing amounts of alumina are being imported.

The upward movement in the energy-output ratio for ammonia production appears to be a result of several factors. First, the industry switched from reciprocal compressors to centrifugal conpressors

Table 4–12. Detailed Energy Use for Industrial Inorganic Chemicals, n.e.c.,
Excluding Diffusion Plants, 1974 and 1976

		Alumina	Ammonia	Other Inorganic Chemicals	Total
1974	Energy use (trillion Btu)	129	260	288	677
	Energy-output ratio (1000 Btu per 1974 dollar)	445	650	108	203
1976	Energy use (trillion Btu)	113	300	221	634
	Energy-output ratio (1000 Btu per 1974 dollar)	452	714	81	186

Source: Table 4-11; *Annual Survey of Manufactures*; Aluminum Association;
The Fertilizer Institute.

beginning in 1965. Energy savings from this switch were probably
slowing by 1972 or 1973 as the early 1970s saw overcapacity and a
sharp dip in new construction. Second, the ammonia shortages of
1974 and 1975 forced plants to run at greater than normal levels. As
a consequence, maintenance was delayed and inefficiencies developed. Third, the natural gas shortage has pushed the industry toward
greater fuel oil use (although the industry is a priority customer for
natural. gas). Finally, environmental energy costs are likely to have
risen.

The acceleration of decline in the energy-output ratio for the industry excluding ammonia and alumina is thus very substantial—an
average drop of more than 10 percent annually. The decrease is, in
fact, sufficiently large as to be suspicious, and we would anticipate
some small revisions in the Census Bureau data either for 1974 or
1976.

Conclusions

Energy use in the industrial inorganic chemical, n.e.c., industry
including diffusion plants increased on a per unit basis between 1974
and 1976. This increase was due to a very large rate of increase in
diffusion plant output, compounded by a small increase in the
energy-output ratio of the diffusion plants.

Energy savings in the inorganic chemical industry excluding diffusion plants have accelerated since 1967, particularly from 1974 to
1976. This acceleration has been due to both slower rates of growth
of ammonia and alumina production—two energy-intensive products—and to increased energy savings in other inorganic chemicals.

PETROLEUM REFINING (SIC 2911)

The petroleum refining industry in 1976 used 3.1 quad of energy for heat and power, including both purchased and self-generated sources of energy. This represents more than 10 percent of all manufacturing energy use and about 4 percent of all U.S. energy use. It is the second largest energy-using industry.

The products of the petroleum refineries are for the most part fuels. Gasoline is by far the most important, but fuels for jet planes (jet fuel), home heating and diesel engines (distillate fuel oils), and industrial heating (residual fuel oil) are also produced in large quantities. Nonfuel petroleum products include petrochemicals (from which synthetic fabrics and plastics are made), asphalt, lubricants (including grease and motor oil), and waxes.

Description of the Industry

Petroleum refineries refine crude petroleum and natural gas liquids. Crude petroleum (or crude oil) is not a uniform substance but a mixture of chemicals. It is composed almost entirely of hydrocarbons (chemicals composed of carbon and hydrogen) but is often contaminated by sulfur and other undesirable elements. The unusual chemical properties of carbon and hydrogen produce a wide spectrum of molecules in crude oil, ranging from short, volatile molecules like propane and butane to the long, heavy strands that make up tars and paraffins. These different molecules have different boiling points and burn in different ways.

In order to turn the crude oil with its mixture of substances into specific fuels or nonfuel products, the oil must first be separated into "fractions," groupings of hydrocarbons with common characteristics. This is done by distillation—carefully controlled boiling that separates molecules by weight. These fractions are then further processed. Since heavier molecules are generally less valuable than lighter ones, the heaviest fractions are cracked—the long chains are broken into shorter ones—by a variety of processes. A number of further techniques are used to rearrange the molecules and combine them to produce uniform final products—lead-free gasoline with 90 octane, residual fuel oil with 0.1 percent sulfur content, and so on.

Energy and the Petroleum Refining Industry

In addition to being a major user of energy, the petroleum refining industry is America's most important direct source of fuels. The industry supplies roughly 40 percent of the energy used in the United States. This dual role in the energy economy has had contradictory effects on energy use in the industry. On the one hand, since the industry uses close to 700 thousand Btu per processed barrel (each barrel contains 5.8 million Btu), it has a very strong economic incentive to save energy as energy prices rise.

On the other hand, the period of rising energy prices has also been a period of unusual economic conditions and new and difficult production conditions. Since 1971, the industry has faced stricter environmental regulation, requirements to remove lead from gasoline and sulfur from residual oil, the Arab embargo, price controls, rationing, gasoline and fuel oil shortages (leading to government-mandated production of fuel oil), crude allocation, and threatened dismemberment. These external influences either directly required more energy use or produced production inefficiencies that led to greater energy use.

As a consequence, although there were steady efforts made to save energy, the net result through 1974 was an increase in energy use per constant dollar of output. Although 1975 and 1976 could not be considered normal years for the petroleum industry, they were quiet by comparison with events in 1974. Given this modicum of breathing room, the industry reduced energy use per unit of output by more than 3 percent in both 1975 and 1976.

Measuring Energy and Output

Because the primary input of the petroleum refining industry is itself an energy material, substantial amounts of self-generated energy are used in the industry. Self-generated energy materials include refinery gases, petroleum coke, and fuel oil. Since these are derived from a conventional energy source, the following discussion invariably includes them as energy sources. Fortunately, the Bureau of Mines (BOM) publishes an annual series of these data that goes back several decades.

An additional source of energy is heat from exothermic (heat-producing) process reactions. While the industry has captured some of this energy, substantial losses are the general rule. Since this is energy being removed from a conventional energy source, estimation of net energy loss would be quite valuable. Gordian Associates estimates a net available energy loss of roughly 100 thousand Btu per barrel at a typical refinery [1]. These types of calculations, which are based on engineering estimates of model refinery configurations, do not necessarily tell us much about actual practices. A superior method would calculate actual Btu entering all refineries as crude oil and other energy and subtract Btu leaving the refinery as products. This would be an extraordinarily time-consuming data-collection effort. However, we believe that if such statistics existed, they would generally show persistent improvement in the capturing of reaction heat. Because of the lack of data, we have in general ignored this factor in this book.

Output of the petroleum refining industry is often measured in crude runs to stills (barrels of crude oil processed in crude distillation units). Actually, this is an *input* measure rather than an *output* measure. If each input barrel of crude resulted in an equal dollar output of finished products, over time the input measure would be a useful substitute for output. However, for most of the post-war period, output has shifted toward more valuable products—more gasoline with higher octane out of each barrel, for example. Thus for most of the period, crude runs underestimate actual output growth as measured in an output index weighted by real prices. For the period 1947–72, this difference averaged about 1 percent per year.

This trend was completely reversed after 1972 when production of the least valued product, residual fuel oil, began increasing (Table 4–13). As a consequence, the Census Bureau-FRB index increased by 9.4 percent between 1972 and 1976, and crude runs to stills increased by 14.7 percent (Table 4–14).

Historical Developments, 1947–67

Given the very high energy intensity of the petroleum refining industry and the fact that only a small proportion of these costs appear to be theoretically necessary for the refining process, one would expect that systematic savings of energy would occur over time. This would have been the case if the output mix had remained constant. It has been estimated by W.L. Nelson that the energy requirements for a fixed output have dropped between 1950 and 1970 at an annual rate of 2.0 percent [2].

However, the output mix has changed considerably over this period. The major change has been a shift from production of heavy (residual) oils valued at 5 cents a gallon in 1967 into gasoline (12 cents), jet fuel (10 cents), or distillate oil (9 cents). There has also been an increasing output of ethylene (18 cents) and other petrochemical feedstocks. This has required additional processing of the heavy and dense molecules of residual fuel oil, "cracking" them into shorter units by thermal, catalytic, and hydrogen methods and then rearranging them into molecules with the most desirable properties via catalytic reforming, alkylation, and other processes.

These additional processes are typically much more energy intensive than the basic distillation operation, which merely separates the

Table 4-13.   Product Mix of the Petroleum Refining Industry, Selected Years, 1967-76

	1967	1971	1972	1973	1974	1975	1976
				(million barrels)			
Gasoline	1838.5	2197.6	2315.8	2398.8	2336.4	2392.7	2516.0
Jet fuel	273.2	304.7	310.0	313.7	305.1	318.0	335.8
Distillate fuel oil	804.4	910.7	962.4	1029.3	973.8	968.4	1070.0
Residual fuel oil	276.0	274.7	292.5	354.6	390.5	451.0	504.0
Kerosene	99.1	86.3	79.0	79.4	56.6	55.5	55.6
Liquefied refinery gas; petrochemical feedstocks and ethane	198.9	241.2	254.4	269.4	258.1	235.6	289.0
All other	478.1	546.7	573.0	611.5	606.3	573.8	601.5
Total output	3968.2	4561.9	4787.7	5056.7	4926.8	4995.0	5371.9
Total crude petroleum input	3582.6	4087.8	4280.9	4537.3	4428.7	4541.4	4910.2
				(percent of total)			
Gasoline	46.3	48.2	48.4	47.4	47.4	47.9	46.8
Jet fuel	6.9	6.7	6.5	6.2	6.2	6.4	6.3
Distillate fuel oil	20.3	19.9	20.1	20.4	19.8	19.4	19.9
Residual fuel oil	7.0	6.0	6.1	7.0	7.9	9.0	9.4
Kerosene	2.5	1.9	1.6	1.6	1.2	1.1	1.0
Liquefied refinery gas; petrochemical feedstocks and ethane	5.0	5.3	5.3	5.3	5.2	4.7	5.4
All other	12.0	12.0	12.0	12.1	12.3	11.5	11.2
Total output	100.0	100.0	100.0	100.0	100.0	100.0	100.0

Source: Bureau of Mines, *Minerals Yearbook*, various years.

Table 4-14. Comparison of Output Indexes, Selected Years, 1967-76
(1967 = 100)

	Basis	
	Crude Runs to Stills	*Census Bureau-FRB Index*
1967	100.0	100.0
1971	114.1	112.0
1972	119.5	115.9
1974	123.6	118.5
1975	126.8	118.5
1976	137.1	126.8

Source: Bureau of Mines, *Mineral Industry Surveys*, "Petroleum Statement, Annual," various years; *Census of Manufactures; Indexes of Industrial Production.*

crude oil into lighter or heavier fractions. Crude distillation uses less than 15 percent of the total energy required by a modern refinery.

As a consequence, the average amount of energy absorbed in processing a barrel of crude oil actually rose over 1947-67 period, although at a relatively modest 0.3 percent annual rate. These increases were concentrated in the earlier part of the period during which gasoline production and octane ratings were increasing most rapidly. From 1947 to 1961, energy used per barrel of crude increased at a 1 percent rate annually; it fell at about a 1 percent rate from 1961 to 1967 (Table 4-15).

However, our basic standard for comparison is not how much raw material was processed but rather how much was produced for each unit of energy. From 1947 to 1967, the average output of U.S. refineries contained a larger and larger proportion of more valuable products. Measured in these terms, the picture changes considerably. For the whole period, energy used per dollar dropped at an average rate of 0.6 percent per year (Table 4-16). From 1962 to 1967, the rate picked up to the 2 percent annual decline that Nelson believes represents the true rate of technical progress of the industry.

Energy Conservation, 1967-76

Energy savings between 1967 and 1976 did not sustain the high rates attained in the mid-1960s. Between 1967 and 1974, the ratio of energy use to real output rose at an average rate of 0.4 percent (Table 4-17). Substantial energy savings were concentrated in the last two years, 1975 and 1976, when the energy-output ratio fell at a 3.4 percent annual rate. For the entire period, the rate of energy savings was only 0.4 percent.

Table 4–15. Energy Use and Crude Runs, 1947–76

	Gross Energy (trillion Btu)	Crude Runs to Stills (million barrels)	Gross Energy per Crude Run to Stills $(1) \div (2)$ (1,000 Btu per barrel)
	(1)	(2)	(3)
1947	1,241	1852.2	670
1948	1,332	2048.3	650
1949	1,373	1944.2	706
1950	1,416	2094.9	676
1951	1,571	2370.4	663
1952	1,569	2441.2	643
1953	1,684	2554.9	659
1954	1,765	2539.6	695
1955	1,946	2730.2	713
1956	2,088	2905.1	719
1957	2,295	2890.4	794
1958	2,042	2789.4	732
1959	2,145	2917.7	735
1960	2,272	2952.5	770
1961	2,306	2987.2	772
1962	2,347	3069.6	765
1963	2,357	3170.7	743
1964	2,435	3223.3	755
1965	2,497	3300.8	756
1966	2,522	3447.2	732
1967	2,587	3582.6	722
1968	2,748	3774.4	728
1969	2,854	3879.6	736
1970	2,900	3967.5	731
1971	2,922	4087.8	715
1972	3,070	4280.9	717
1973	3,222	4537.3	710
1974	3,149	4428.7	711
1975	3,042	4541.4	670
1976	3,148	4910.2	641

Source: *Mineral Industry Surveys.*

The ratio of energy to input fell much further. From 1967 to 1976, the ratio of energy used for crude runs to stills fell at a 1.3 percent annual rate. This is due to the fact that less gasoline and more residual fuel oil was being made—therefore, each barrel required less processing.

The low rate of energy savings between 1967 and 1971 was probably primarily due to moderation in demand. Because of both a slowing of general economic growth and warm winters in 1970 and 1971, capacity utilization was 5 percent below the 1967 rate (Table 4–18). In addition, pollution regulation probably increased energy requirements by approximately 1 percent.

Table 4–16. Energy Use and Value Added in Petroleum Refining, Selected Years, 1947–67

	Gross Energy (trillion Btu)	Value Added (billion 1974 dollars)	Energy-Output Ratio (1,000 Btu per 1974 dollars)
	(1)	(2)	(3)
1947	1,241	2.98	416
1954	1,765	4.19	421
1958	2,042	4.88	418
1962	2,347	5.71	407
1967	2,587	7.06	366

Source: *Mineral Industry Surveys; Census of Manufactures.*

However, the situation was quite different after 1971. The period from 1972 to the present can be seen as an epochal transition in world energy use—but this was only the greatest of the influences on the petroleum refining industry.

Product mix changes in this period were greatly affected by environmental regulation. Refiners were required to produce unleaded gasoline as of the middle of 1974 as part of the Environmental Protection Agency's regulation of auto emissions. For a given octane level, producing unleaded gasoline requires substantially more energy than leaded gasoline. As natural gas supplies leveled off, demand for low sulfur fuel oils increased, which in turn required more desulfurization, particularly hydrogen processing. This also increased energy use.

A new regulatory tone was set by the wage and price controls adopted in 1971 and 1972 and the 1972 amendments on water pollution abatement. Moreover, as domestic demand outraced crude oil supply, increasing amounts of crude oil had to be imported. Some of these oils were much more heavy and sulphurous (sour) than the domestic sweet crudes used previously. These oils required greater processing to reach the same end' products—more cracking and deeper desulphurization.

Then in late 1973, the Arab oil embargo struck. Suddenly the major question became how to get oil at all. Massive federal regulation ensued with every phase of industry activity being affected. Perhaps the most glaring example of the inefficiencies caused by the regulations was the crude oil allocation program. This and other measures designed to protect small refineries (and thus preserve both competition and equity) made bringing old refineries out of retirement

Table 4–17. Energy Use and Value Added, Petroleum Refining, 1967–76

	Gross Energy (trillion Btu)	Value Added (billion 1974 dollars)	Energy-Output Ratio (1,000 Btu per 1974 dollar)	Annual Rate of Change from Previous Period (percent)
1967	2,587	7.06	366	
1971	2,922	7.91	369	+ 0.2
1972	3,070	8.18	375	+ 1.6
1973	3,222	8.62	374	− 0.4
1974	3,149	8.36	377	+ 0.8
1975	3,042	8.36	364	− 3.4
1976	3,148	8.95	352	− 3.3

Source: Mineral Industry Surveys; Table 4–15; Annual Survey of Manufactures.

Table 4-18. Capacity Utilization in Petroleum Refining, 1967-76

	Operable and Operating Capacity (1,000 barrels per day)	Total Crude Runs (1,000 barrels per day)	Capacity Utilization (percent)
1967	10,658	9,815	92
1971	12,860	11,200	87
1972	13,292	11,696	88
1973	13,671	12,431	91
1974	14,362	12,133	84
1975	14,961	12,408	83
1976	15,237	13,416	88

Source: American Petroleum Institute: *Mineral Industry Surveys.*

profitable. Since old refineries are retired precisely because they are inefficient, bringing them back into production meant less efficient production—and greater energy use. In 1973 and 1974, according to the *Oil and Gas Journal*, capacity rose by roughly 11 percent without a single new grass roots facility being built. As a result of all this, the energy-output ratio rose 0.7 percent from 1971 to 1974.

Although 1975 was scarcely normal, the changes were not nearly as dizzying. With the embargo in the past, refinery output moved up despite the worst recession since the 1930s. Capacity grew very slowly, but this was an improvement over capacity created by bringing inefficient plants out of retirement. The upshot was that the energy-output ratio turned around and dropped by 3.4 percent. This downward trend continued through 1976 when the ratio dropped another 3.4 percent.

Despite continuing regulation, 1976 was a year of significant progress. Production rose by over 7 percent, and capacity rose by nearly 1.1 million barrels. In sharp contrast to the 1973-74 period, the bulk of this growth was in new plants and major additions to capacity. Half of it was accounted for by the new Marathon Oil refinery in Louisiana (200 thousand barrels per day) and two major expansions in California (175 thousand barrels each).

During the first year that a plant or major addition is in place, it normally operates at somewhat less than full efficiency. This is due to start-up costs and the learning time required for any highly complex machinery. Thus, these new investments should lay the basis for even greater energy savings in 1977 and succeeding years.

Comparisons with American Petroleum
Institute Energy Data

Table 4-19 shows a comparison of two sources of data on energy: data reported by the American Petroleum Institute (API)[e] to the federal government as part of VIECP and data published by the BOM. The most striking difference between the two series is in the statistics for natural gas and refinery gas. The BOM reported 240 trillion more Btu of gas than did API. Since this figure represents nearly one-twelfth of total energy consumed by the industry (and nearly 1 percent of energy absorbed by U.S. manufacturing), the difference is significant. We encourage efforts by the two groups to reconcile the two sources.

We have chosen to use the BOM data but *not* because we have any basis for judging whether their data for 1976 are necessarily superior in accuracy to the data of API. Instead, we chose the BOM data because it provides a superior basis for historical comparison. API does report 1972 energy data, but they are based on an incomplete sample (77 percent of estimated energy consumption). Obviously the BOM—whose reporting system goes back decades—provides a more comprehensive basis for analysis.

Energy for Pollution Abatement

The API surveys include estimates of energy required for the processing of liquid, gaseous, and solid wastes. These are reported as

Table 4-19. Comparison of Bureau of Mines and API Energy data, 1972
(trillions of Btu)

	API	Mines
Crude oil	2.8	.6
Distillate oil	27.0	27.0
Residual oil	278.5	296.6
Liquified petroleum gas	42.4	39.1
Natural gas	738.2	938.2
Refinery gas	1100.0	1141.4
Petroleum coke	433.1	395.6
Coal	4.7	4.2
Purchased steam	38.2	39.8
Purchased electricity	214.5	265.6
	2879.4	3148.1

Note: Minor differences may exist in conversion factors.
Source: American Petroleum Institute; *Mineral Industry Surveys.*

[e]API are also the initials of the American Paper Institute (Chapter 3). In this chapter, the initials invariably refer to the American Petroleum Institute.

increments over the 1972 baseline adopted for the surveys. Coverage varies somewhat from survey to survey, but each represents more than 80 percent of the industry.

These surveys estimate that direct pollution abatement caused energy consumption to be roughly 0.2 percent higher in late 1974 compared with 1972 (per input barrel). The estimate for 1975 was 0.3 percent and for 1976 0.5 percent. It appears that for the period 1974–76, incremental energy costs for direct pollution abatement averaged between 0.1 and 0.2 percent annually.

Conclusions

Between 1974 and 1976, the petroleum refining industry saved energy at a pace much greater than previously. From 1974 to 1976, the energy-output ratio declined at a rate greater than 3 percent annually. This compares very favorably with the period from 1947 to 1967 when the ratio fell less than 1 percent annually.

Between 1967 and 1974, the ratio of energy use to real output rose. The major reasons for this reversal in the general trend of energy savings appear to be the unusual conditions under which the industry operated in the period from 1971 to 1974. These conditions included extensive governmental regulation affecting all phases of refinery activity, such as crude oil inputs, pollution abatement, and output mix.

NOTES TO CHAPTER 4

1. Gordian Associates, *The Data Base*, vol. 2, NTIS, 1974.
2. *Oil and Gas Journal*, December 30, 1974.

Chapter 5

Durable Goods Industries

Cement, steel, and aluminum are combined in this chapter because they are more subject to cyclical variability than nondurable goods. The unusually wide swings in business activity experienced during the 1967–76 period affected the three industries covered in this chapter to a greater degree than the five industries discussed in Chapters 3 and 4.

For this book, the most important feature of these wide variations in production is their effects on energy use. In the aluminum industry, for example, output rose 50 percent in seven years, and then fell 20 percent in the following year (1975). It is difficult to determine the extent of energy conservation in an industry subject to such fluctuations in output. The findings for these industries are therefore more tentative than those for the industries discussed in Chapters 3 and 4.

HYDRAULIC CEMENT (SIC 3241)

Although among the eight industries covered in this book the cement industry ranks last as an energy user, it is second in energy intensiveness as measured by Btu per dollar of value added. The industry is very closely tied to the volume of construction activity; for this reason it often exhibits cyclical behavior different from swings in general business activity. During the 1960s, for example, the industry had excess capacity, grew slowly, and earnings of the companies in the industry declined steadily [1].

Products and Processes

Portland cement accounts for 94 percent of the output of this industry. Nearly all the other products are other types of cement. The 1972 Census of Manufactures reported 199 plants in the industry in 1972, owned by 75 companies.[a] Portland cement is produced by grinding limestone, mixing it with small amounts of other materials (clay, shale, sand, etc.), and then firing the mixutre in a rotary kiln to produce clinker—marble-sized pellets—which is then ground and mixed with gypsum to produce the final product.

Although most cement in the United States (56 percent in 1975), is produced by the wet process, the proportion is falling. In the wet process, water is blended with the raw materials before kiln-firing; this step is omitted in the dry process.

Energy and Output Data

Annual data are important for this industry because of the wide swings in output. The Bureau of Mines (BOM) publishes a wealth of information about the industry, including annual data on output of clinker and cement and energy use by source; all of these are cross-classified by production process [2]. Capacity figures are also published by the BOM, but the concept varies over time: from 1947 to 1969, only finish grinding capacity was available; from 1970 to 1972, only clinker capacity was given; from 1973, both finish grinding and clinker capacity were published.

For the period through 1969, imports of clinker were small—less than 0.5 percent of cement production. Thereafter, imports of clinker rose rapidly, reaching a peak in 1973 of 3.3 percent of cement production, and then declining to 1.4 percent in 1976 (Table 5-1).

Clinker is an intermediate product in cement manufacture, and its production is very energy intensive. Approximately 81 percent of industry energy consumption is accounted for by clinker burning and another 11 percent in preceding operations (drying and crushing raw materials, etc.). Only 8 percent is used in grinding cement [3]. Clinker production is more closely related to energy use than cement production for this reason.

When significant amounts of clinker are imported, however, the energy used in grinding this imported material into a finished product rises in importance. A composite measure of output was constructed, using a weight of 0.92 for clinker production and 0.08 for cement production, derived from the preceding figures. The weights

---

[a]Other sources, such as the Bureau of Mines, show somewhat fewer plants and companies.

Table 5–1. Output Characteristics of the Cement Industry, 1967–76

| | Production | | Capacity | | Capacity Utilization | | Clinker Imports |
| | Clinker | Cement | Clinker | Cement | Clinker | Cement | |
	(thousand tons)		(thousand tons)		(percent)		(thousand tons)
1967	71,172	69,447	NA[a]	95,683	NA	72.6	25
1968	74,828	74,243	NA	95,703	NA	77.6	29
1969	75,628	75,125	NA	95,370	NA	78.7	114
1970	73,990	73,168	83,675	NA	88.4	NA	402
1971	75,231	73,007	85,791	NA	87.7	NA	728
1972	77,378	80,744	85,399	NA	90.6	NA	1,691
1973	78,212	83,551	86,882	100,413	90.0	80.2	2,743
1974	77,978	79,486	90,874	106,223	85.8	74.8	1,829
1975	64,539	66,796	92,264	106,111	70.0	62.9	1,207
1976	68,579	71,227	85,419	104,146	80.3	68.4	962

[a]NA means not available.

Source: *Minerals Yearbook*, various issues; telephone conversations with Bureau of Mines officials.

used should vary over time as technology changes at different rates for various operations. The weights of 0.92 for clinker and 0.08 for cement are reasonably accurate for the 1970–76 period, however, and clinker production did not differ appreciably from cement production before 1970 when clinker imports first became important. The fixed weights seem, therefore, to be a valid approximation for the entire period, 1947–76.[b]

The capacity problem is somewhat more complex. We have been told by industry representatives that there are significant variations in clinker inventories, and thus finish grinding capacity may not be indicative of industry capacity utilization, particularly as it affects energy use. We are, therefore, confined to the post-1970 period for analysis of the effect of capacity on energy use per ton of production. Output is measured in equivalent tons, computed using the weighting method described above. Energy is measured in Btu, derived from the quantities of fuels reported in the BOM *Minerals Yearbook.*

Energy Conservation, 1947–76

Energy use per equivalent ton of output declined at an average rate of 0.7 percent per year over the twenty-nine-year period from 1947 to 1976 (Table 5–2). There was a significant slowing in energy conservation after 1967, however. The annual average rates of change of the energy-output ratio in the two subperiods were −1.0 percent from 1947 to 1967 and zero from 1967 to 1973. The principal explanation for this pattern appears to be the slowdown in growth of the industry. Production grew at a 3.6 percent rate from 1947 to 1967 but at only a 1.7 percent rate from 1967 to the peak year of 1973.

In industries with long-lived capital, such as the cement industry, energy saving per unit of production comes mainly through new plant and equipment that embodies the latest technology. This investment takes place at a more rapid rate when output is expanding briskly than when output growth is slow.

From 1974 to 1975, capacity utilization (of clinker) dropped sharply, and the energy-output ratio rose. In 1976, capacity utilization recovered much of the loss of the preceding year, and the energy-output ratio declined substantially. For the 1973–76 period,

---

[b]The Census Bureau index of production for hydraulic cement rose 15 percent from 1967 to 1972. This agrees more closely with cement production (+16 percent) than with the composite measure (+ 9 percent). The composite measure is a better indicator of productive activity in the industry, however, since it is less affected by the rise in imports of semifinished materials.

Table 5-2. Energy Use, by Source, and Output in the Cement Industry, 1947–76

	Gross Energy (trillion Btu)					Production in "Equivalent Tons" (thousands)	Energy-Output Ratio (million Btu per ton)
	Coal	Oil	Natural Gas	Purchased Electricity	Total		
	(1)	(2)	(3)	(4)	(5)	(6)	(7)
1947	190.1	28.2	64.5	35.9	318.7	35,252	9.041
1948	204.9	28.4	74.9	40.6	348.7	38,987	8.944
1949	191.3	28.0	86.5	40.4	346.2	39,773	8.704
1950	190.2	32.1	99.2	40.4	361.9	42,624	8.490
1951	204.2	38.7	104.9	43.8	391.7	46,659	8.394
1952	193.4	39.7	114.1	45.2	392.3	47,127	8.324
1953	200.3	41.4	119.6	47.1	408.3	49,818	8.196
1954	194.6	40.2	128.7	44.4	407.9	51,129	7.977
1955	209.0	51.9	134.2	46.9	442.0	56,288	7.853
1956	222.0	48.3	147.2	51.9	469.4	60,094	7.812
1957	212.0	33.0	149.2	51.4	445.7	57,125	7.803
1958	201.8	27.3	168.5	54.3	451.9	58,787	7.687
1959	207.6	27.5	194.5	59.0	488.6	64,029	7.631
1960	200.4	24.6	175.2	59.8	460.0	60,928	7.550
1961	186.0	23.9	184.1	62.3	456.3	60,812	7.504
1962	189.4	24.5	191.7	65.0	470.6	63,297	7.434
1963	199.3	24.6	202.6	69.8	496.2	66,572	7.453
1964	211.3	26.4	205.9	75.5	518.9	69,915	7.423
1965	218.8	27.2	202.7	77.7	526.4	70,481	7.469
1966	223.6	23.9	208.1	83.1	538.6	73,170	7.361
1967	217.9	30.2	199.8	82.6	530.5	71,034	7.468
1968	227.7	35.2	207.2	86.7	556.8	74,781	7.445
1969	219.9	37.0	205.5	90.3	552.8	75,588	7.313
1970	170.8	61.2	216.3	91.6	559.8	73,924	7.573
1971	172.0	65.6	224.6	96.7	558.9	75,374	7.415
1972	175.8	74.6	228.0	101.7	580.1	77,648	7.470
1973	178.8	81.7	221.6	105.7	587.8	78,639	7.474
1974	193.9	62.9	212.8	105.2	574.7	78,099	7.359
1975	181.3	44.7	162.9	92.2	481.1	64,720	7.433
1976	222.5	46.4	135.0	98.6	502.5	68,791	7.304

Source: *Minerals Yearbook*, various issues; telephone conversations with Bureau of Mines officials.

the average decline in the energy-output ratio was 0.8 percent, indicating an improvement over the lackluster pattern of the preceding six years. This improvement is all the more striking since it occurred when output was shrinking.

Process Change

The wet process uses substantially more energy per barrel than the dry process (Table 5−3). This results from the need to evaporate the water added in the blending process. As a result of cost pressures, the industry is shifting from the wet to the dry process—the proportion of output produced by the wet process fell from 59 percent in 1967 to 55 percent in 1976. This shift was brought about by the addition of new and expansion of old dry process plants.

An analysis of the 1967−76 data reveals, however, that nearly all the gains in energy conservation have been won by improvements within each process rather than by the rise in the importance of the dry process, which uses less energy. For example, if the 1976 output had been produced with the same proportions of wet and dry process plants that prevailed in 1966 (59 percent and 41 percent, rather than 55 percent and 45 percent), total energy use would have been less than 1 percent higher in 1976 than the actual amount.

Alternative Estimates of Energy Saving

The Portland Cement Association (PCA) has reported to the federal government under VIECP for the period since 1972. The coverage of these reports is less extensive than that of the BOM *Minerals Yearbook*.[c] As a result, energy use and production figures from the two sources move somewhat differently.

From 1972 to 1976, the PCA report showed a 6.7 percent decline in energy use per equivalent ton while the figures in Table 5−2 indicate only a 2.2 percent decline. Part of the difference between the two sets of figures results from the conversion factor used for purchased electricity. The PCA, following the federal guidelines, converts purchased electricity at 3,412 Btu per kwh while we use the heat rate. Since purchased electricity rose in importance as an energy source from 1974 to 1976 (and self-generated electricity fell absolutely), the energy total in the PCA report fell more than our measure from 1972 to 1976.

After elimination of the effect of the electricity conversion factor, a substantial difference still remains: the PCA output (derived in the

---

[c]Puerto Rico is included in the *Minerals Yearbook* statistics but excluded from the PCA reports, for example.

Table 5-3. Energy Use, by Source, and Output in the Cement Industry, by Production Process, 1966-76

Production Process and Year	Gross Energy (trillion Btu)					Production in "Equivalent Tons" (thousands)	Energy Output Ratio (5)÷(6) (million Btu per ton)
	Coal	Oil	Natural Gas	Purchased Electricity	Total		
	(1)	(2)	(3)	(4)	(5)	(6)	(7)
Wet							
1966	128.1	20.8	144.4	51.7	345.0	43,497	7.931
1967	125.9	26.6	141.8	52.4	346.8	43,296	8.009
1968	133.0	30.4	143.4	54.8	361.6	45,701	7.911
1969	133.0	31.1	142.3	57.6	363.9	46,106	7.893
1970	104.1	45.7	146.3	56.3	352.3	44,624	7.895
1971	97.5	49.7	149.0	58.3	354.5	45,101	7.861
1972	99.6	53.2	150.6	59.5	363.0	45,903	7.907
1973	102.3	59.4	146.2	61.6	369.4	46,200	7.995
1974	104.8	47.9	142.0	59.7	354.3	45,037	7.868
1975	101.0	33.9	107.2	51.5	293.5	36,508	8.039
1976	121.0	33.6	90.0	53.5	298.0	38,122	7.816
Dry[a]							
1966	95.5	3.1	63.7	31.3	193.6	29,674	6.524
1967	91.9	3.6	58.0	30.2	183.7	27,738	6.622
1968	94.8	4.7	63.8	31.9	195.2	29,081	6.712
1969	86.9	6.0	63.3	32.7	188.9	29,482	6.406
1970	86.7	15.5	70.0	35.4	207.5	29,300	7.083
1971	74.5	15.9	75.6	38.4	204.4	30,272	6.750
1972	76.2	21.4	77.4	42.2	217.1	31,744	6.839
1973	76.5	22.4	75.4	44.1	218.4	32,440	6.732
1974	89.1	15.0	70.8	45.5	220.4	33,063	6.665
1975	80.3	10.8	55.7	40.7	187.6	28,212	6.649
1976	101.5	12.8	45.0	45.1	204.5	30,669	6.667

[a]Includes a small number of plants using both wet and dry processes.

Source: *Minerals Yearbook*, various issues; telephone conversations with officials at the Bureau of Mines.

same manner as our own) shows a smaller decline than our own from 1972 to 1976 while the PCA energy total shows a greater decline. At least part of this puzzling result is attributable to coverage and to different conversion factors used for other fuels.

Components of Energy Saving

Clinker capacity of the industry rose 10 percent from 1970 to 1975. This is, of course, a net change because many old plants were retired during the period (primarily wet process). This expansion and turnover contributed the bulk of the energy savings over the period. New plants, and additions to existing plants, are nearly always larger than the existing average size, and large plants (and large kilns and large grinding mills) use less energy per ton than smaller ones [4]. This is but one aspect of the broader concept of economies of scale, of course, but it is the aspect that concerns us particularly in this report. The annual narrative in the BOM *Minerals Yearbook* is replete with cases of new kilns, grinding mills, and entire plants that are larger than both the industry average and the facilities they replace.

A second source of energy savings is add-on equipment for existing plants. One of the most important types of add-on equipment is the preheater. About 80 percent of the energy used within the cement manufacturing plant is in the kiln where the clinker is "burned" or calcined. Substantial savings can be achieved by reusing the exit gases from the kiln to preheat the ground limestone and other minerals, to evaporate the moisture in the materials, and to partly calcine the raw material. The PCA reports that the number of preheaters rose from nineteen at the end of 1970 to fifty-four or fifty-five at the end of 1976. Practically all new plants are equipped with preheaters. Improved insulation of existing kilns is another way in which energy may be conserved.

Fuel Substitution

The wet process plants use more oil and natural gas and less coal and purchased electricity per ton of cement than dry process plants (Table 5-3). From 1966 to 1971, there was a movement away from coal in both processes toward fuel oil and purchased electricity. Electricity is used in grinding both limestone and clinker, and comparatively little is generated within the industry.

Since 1971, there has been some movement back to coal as well as a decline in natural gas, and these trends will continue, according to the BOM annual narrative, because of supply difficulties and rising fuel prices.

Energy Use in Pollution Control

The PCA attributes much of the rise in electricity use in recent years to the operation of precipitators and bag houses to control air pollution. Purchased electricity per ton of product rose 9 percent in the wet process and 16 percent in the dry process from 1971 to 1976. In the earlier year, there were apparently very few air pollution control facilities in place. We estimate, based on this movement in electricity use, that total energy use was at most 2 percent higher in 1976 because of pollution control.

Conclusions

During the 1967–73 period, there were little or no energy savings in the cement industry. The following three years, 1973–76, saw a significant improvement, however, despite the fact that output declined by more than 12 percent.

Turnover of old plant and equipment accounted for most of the recent energy conservation. Improvement in existing capital by addition of energy-saving devices and better insulation also contributed to the savings. A small amount of savings resulted from the continuing shift from the wet to the dry manufacturing process. Air pollution control was a restraining element, however, offsetting part of the gains in energy conservation obtained by these methods.

BLAST FURNACES AND STEEL MILLS (SIC 3312)

Steel is the largest energy user in manufacturing; the industry consumed more than 4 percent of the entire nation's energy in 1976. Because of its size, therefore, modest percentage gains in energy conservation in the industry bulk large in the performance of the manufacturing total. This is an industry of complex structure that is characterized by slow growth, large cyclical swings in output, and difficulties with competing imports. These elements make the winning of energy savings a difficult task.

Description of the Industry

The principal raw materials used are iron ore, coal, limestone, and steel scrap. More than 90 percent of the coal is processed into coke in coke ovens, which is then combined with iron ore and limestone in blast furnaces to produce pig iron. The latter is then combined with steel scrap in steel-melting furnaces to produce raw steel. The raw steel is then converted into basic shapes (plates, sheets, strips, etc.) in rolling mills.

Three types of steel-melting furnaces are in use in the United States: basic oxygen (BOF), open hearth (OH), and electric (EF). The electric furnace differs from the description in the preceding paragraph in that it uses steel scrap almost exclusively instead of pig iron in making steel.

The *1972 Census of Manufactures* reported 245 companies with 364 plants in the industry, but most of these were comparatively small. Thirty-six fully integrated plants (coke ovens through rolling mills) with 1,000 or more employees accounted for more than half of the output of the industry, and forty-one partially integrated plants in the same size range accounted for another one-fourth of the output. That is, seventy-seven large integrated or partially integrated plants produced more than three-fourths of the output (83 percent of value added) of this large industry. For this reason, we shall concentrate on integrated steel mills in our discussion of the industry.

Growth and Cyclical Variation

From 1953 to 1974, production of raw steel grew at an annual average rate of 1.2 percent. Production rose at a more rapid rate from 1953 to 1966 (1.4 percent) than from 1966 to 1974 (1.0 percent). The years chosen in this measurement, 1953, 1966, and 1974, were similar in capacity utilization, and thus the numbers cited provide a reasonably accurate measure of change over a long period. In view of the fact that the output of total manufacturing rose at an average annual rate of 4.1 percent from 1953 to 1974, we see that steel grew only about one-third as fast.

The great cyclical variability in steel is illustrated by the middle line in Figure 5-1. In the major recessions of 1954, 1958, and 1975, production fell sharply. Energy use per ton of raw steel tends to rise when output declines as shown by the third line on the figure.

Measurement of Output and Energy Use

An annual measure of energy use is highly desirable for the steel industry in order to avoid misinterpretation of cyclical patterns for long-term movements. This is a sufficient reason for using data from the American Iron and Steel Institute (AISI) in place of periodic census data. A second reason is the detailed statistics on the use of metallurgical coal included in the AISI publications. This information is not available in census publications, particularly after 1967. Metallurgical coal accounts for well over 50 percent of total energy use by the industry. For these reasons, the following discussion is based largely on AISI data. We now turn to a discussion of possible measures of output and energy use.

Figure 5–1. Energy and Raw Steel Production

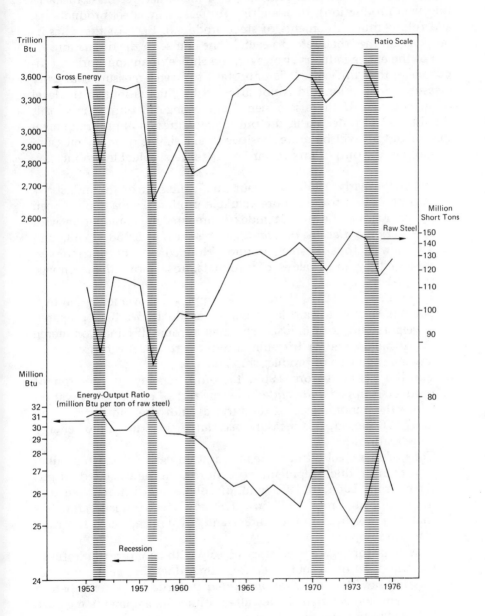

Source: Table 5–4.

To begin, we note that there are measures of output that correspond roughly to the different stages of manufacture: blast furnaces (pig iron production), steel-melting furnaces (raw steel production), and rolling mills (shipments of steel products). Each of these has its purpose, and we shall refer to each in the course of the discussion.

For the entire industry, however, the choice of an appropriate output measure is difficult. It is important to have a measure that truly measures production and is not affected by fluctuations in the level of inventories. If there is a significant change in output mix, this should also be reflected in the output measure; an appropriate measure of output will be more sensitive than a simple quantity measure to shifts in the output mix toward more, or less, valuable products.

1. From the early 1950s to about 1967, the mix of the industry's output shifted toward more valuable products. This may be seen by comparing the output indexes prepared in conjunction with the periodic *Census of Manufactures* (1954, 1958, 1963, and 1967) with tons of steel products shipped; the output indexes, weighted by unit values of products, rose more than tonnage shipped.

2. From 1967 to 1976, there was an opposite movement—the output mix shifted toward less valuable products. We have prepared a weighted index of industry shipments from AISI and Commerce Department data for fifty-nine product groups; this measure rose less than tonnage of product shipments.

3. For the period before 1967, therefore, energy use per ton of products shipped *understated* the true rate of energy conservation (since the denominator of the ratio did not rise enough); for the 1967–76 period, this measure *overstated* the true rate of energy conservation.

4. Tons of raw steel output, measured at an earlier stage in production than product shipments, rose more than the weighted output index from 1954 to 1967 and, of course, much more than tonnage shipped. From 1967 to 1976, raw steel tonnage grew less than the weighted output index and, of course, much less than tonnage shipped.

5. Raw steel tonnage agrees more closely with an index that reflects the true level of output, therefore, than does shipments tonnage. But the raw steel measure *overstated* the true output rise before 1967 and *understated* it thereafter, while the shipments measure behaved in the opposite fashion.

6. Shipments of steel products are affected by inventory change. This is revealed by the ratio of raw steel *production* to product

shipments (unweighted) (Table 5−4). The rise in the ratio from 1953 to about 1967 and decline thereafter are related to the changes in product mix discussed in 1 and 2 above. The year-to-year fluctuations in the ratio demonstrate mainly inventory change as well as some variation in product yield.

7. There remains the possibility that there has been a general improvement since 1967 in product yield that our value-weighted output indexes have not captured. This would help to explain the decline in the ratio of raw steel to shipments shown in Table 5−4 (column [6]). Such a result could have come from a decline in scrappage, for example. If product yield has risen since 1967, the raw steel measure would fail to capture this effect.

From these considerations, we have decided to show two measures of energy conservation for the industry—energy use per ton of raw steel production and energy use per ton of steel product shipments. We shall attempt to guard against misinterpretation resulting from the product mix and inventory change problems discussed in 5 and 7 above.

For the measurement of energy use, there are two common approaches for the steel industry. The Census Bureau classifies coal use for the production of coke as a raw material rather than an energy source. Coke is one of the three essential ingredients in iron production (the others are limestone and iron ore) at present.[d] The reasoning is that coke cannot be replaced by other fuels in the blast furnace. Coal used for steam and other purposes is, of course, treated as fuel in census usage, as are coke and coke oven gas.

The other approach, adopted by the steel industry in its reports to the federal government, is to treat all coal consumed in the industry as fuel. The rationale is that some substitution of other fuels for coke is possible in the blast furnace (injection of oil to improve combustion, for example) and that all of the heat content of metallurgical coal is consumed in the steel industry—that is, the product is not combustible.

We have estimated energy use in the steel industry employing both concepts. The levels differ somewhat—the more inclusive measure is higher—but the relative changes are quite similar. We have adopted the industry concept for this report because it is more simple and because the substitution argument just mentioned is compelling in

[d]At this writing, no significant amount of iron is being produced by direct reduction (an alternative to the blast furnace, which does not use coke). Efforts to overcome technical problems associated with the process are currently underway at several locations.

Table 5-4. Energy, Production, and Shipments in the Steel Industry, 1953-76

	Gross Energy[a] (trillion Btu)	Raw Steel Production (million tons)	Energy to Raw Steel Ratio (1) ÷ (2) (million Btu per ton)	Shipments of Products (million tons)
	(1)	(2)	(3)	(4)
1953	3,459	111.61	30.99	80.15
1954	2,795	88.31	31.65	63.15
1955	3,477	117.04	29.71	84.72
1956	3,425	115.22	29.73	83.25
1957	3,489	112.72	30.95	79.90
1958	2,648	85.26	31.06	59.91
1959	2,750	93.45	29.43	69.38
1960	2,914	99.28	29.35	71.15
1961	2,749	98.01	28.05	66.13
1962	2,781	98.33	28.28	70.55
1963	2,923	109.26	26.75	75.56
1964	3,333	127.08	26.23	84.94
1965	3,477	131.46	26.45	92.67
1966	3,469	134.10	25.87	90.00
1967	3,346	127.21	26.30	83.90
1968	3,406	131.46	25.91	91.86
1969	3,608	141.26	25.54	93.88
1970	3,546	131.51	26.96	90.80
1971	3,246	120.44	26.95	87.04
1972	3,413	133.24	25.62	91.80
1973	3,775	150.80	25.03	111.43
1974	3,743	145.72	25.69	109.47
1975	3,306	116.64	28.34	79.96
1976	3,329	128.00	26.01	89.45

[a]Adjusted for change in coke inventory.

Source: Derived from data reported in American Iron and Steel Institute, *Annual Statistical Report*, various issues; Federal Reserve Board.

our judgment. In the analysis of components of the industry, we shall use by-product fuel measures, such as coke oven gas and blast furnace gas, because these are direct substitutes for other fuels in specific processes.[e]

Energy Conservation, 1953-76

The use of annual data permit us to isolate to a large extent the impact of cyclical fluctuations on energy use per ton of raw steel or product shipments. As shown in column (7) of Table 5-4, the level of capacity utilization, as measured by the FRB index, was similar in 1953, 1955, 1965, 1966, and 1974. As an approximation to the

[e]Data on industry components are incomplete so that it is not possible to disaggregate completely the energy-output ratio for the industry.

Table 5−4. continued

	Energy to Shipments Ratio (1) ÷ (4) (million Btu per ton)	Raw Steel to Shipments Ratio (2) ÷ (4)	Capacity Utilization: Iron and Steel, Subtotal (percent)
	(5)	(6)	(7)
1953	43.16	1.39	95.7
1954	44.26	1.40	72.4
1955	41.04	1.38	95.3
1956	41.14	1.38	90.9
1957	43.67	1.41	86.2
1958	44.20	1.42	63.7
1959	39.64	1.35	69.7
1960	40.96	1.40	71.6
1961	41.57	1.48	69.7
1962	39.42	1.39	71.2
1963	38.68	1.45	78.0
1964	39.24	1.50	88.5
1965	37.52	1.42	94.3
1966	38.54	1.49	94.0
1967	39.88	1.52	84.3
1968	37.08	1.43	84.5
1969	38.43	1.50	90.7
1970	39.05	1.45	83.6
1971	37.29	1.38	75.9
1972	37.18	1.45	85.7
1973	33.88	1.35	98.1
1974	34.19	1.33	94.7
1975	41.35	1.46	74.7
1976	37.22	1.43	80.4

long-term trend in energy conservation, we can examine rates of change between 1953 and 1974. In addition, two subperiods are of interest, 1953−66 and 1966−74. The rates of change shown below are derived from Table 5−4.

	Annual Average Rate of Change Energy Use per Ton of	
	Raw Steel Production	Shipments of Products
1953−74	−0.9	−1.1
1953−66	(−1.3)	(−0.9)
1966−74	(−0.1)	(−1.5)
1974−76	+0.6	+4.5

These figures indicate that the long-term rate of energy conservation in the industry is about 1 percent per year, using either measure.

For the two subperiods, the raw steel and product shipments measures give conflicting results; the raw steel ratio indicates a slowing since the mid-1960s, while the steel shipments ratio indicates an acceleration. For the reasons outlined above, we feel that the raw steel measure is superior and that the rate of energy conservation in the steel industry slowed after the mid-1960s. But the record for the 1966–74 period is not our principal concern in this book. Instead, we seek to determine what has happened in the industry since the 1973 embargo. Here the evidence is unambiguous: energy use has risen regardless of the measure of output used. The principal reason for the rise in energy use per unit of output is the cyclical effect; capacity utilization fell 20 percentage points in 1975, according to the FRB measure, and recovered only 5.7 points in 1976. Another reason was interruptions in the supply of natural gas, forcing steel mills to convert abruptly to other fuels (oil or coal) that were often less energy efficient in a technical sense in specific operations.

Industry Estimates of Energy Conservation

In its reports to the federal government under VIECP, the AISI reports energy use per ton of net shipments, adjusted by the reporting companies for inventory change [5]. The coverage of the companies surveyed by the AISI varies, resulting in different estimates of energy conservation over time. The most recent reports, based on more complete coverage than earlier reports, indicates a 4 percent improvement in energy conservation from 1972 to 1976. This suggests that energy conservation in the industry proceeded at the long-term pace of about 1 percent per year during this period.

Intermediate years were not reported in the most recent reports—only 1972, 1976, and the first half of 1977. In order to examine intervening years, we must resort to a smaller sample, still covering more than 90 percent of the industry, however. Both sets of estimates are shown below. Estimates from the earlier sample imply a smaller gain in energy conservation over the entire four-year period (2.8 percent as opposed to 4.0 percent) and may be less accurate. The year-to-year fluctuations are of interest, however: they suggest that nearly all of the gains in the 1972–76 period were won in 1973 when capacity utilization rose to a very high level; from 1974 to 1976, essentially no change was accomplished. This pattern supports our contention that cyclical effects are often controlling in the determination of energy conservation in this industry.

Million Btu per Net Ton of Shipments[f]

	Latest Sample	Earlier Sample
1972	33.14	32.61
1973	—	31.73
1974	—	31.71
1975	—	33.58
1976	31.80	31.71

Energy Use in Components of the Industry

It is possible to derive at least partial measures of energy use per unit of output in three of the four principal components of the industry: blast furnaces, part of steel-melting furnaces, and heating and annealing furnaces (components of the forming of raw steel into semifinished shapes).

Blast furnaces are by far the largest energy-using component, accounting for more than one-half of total energy use. Reasonably complete data show a slight decline in fuel use per ton of pig iron (Table 5-5). From 1963 to 1976, this ratio fell at a 0.6 percent average annual rate. Coke use per ton of pig iron fell substantially, but this was partially offset by increases in the use of fuel oil and tar and pitch (coke oven by-products).[g]

Steel-melting furnaces use comparatively little energy: only about 10 percent of blast furnace use [6]. There has been a sharp decline in fuel use per ton of raw steel by OH and BOF furnaces since 1963 (Table 5-5). This ratio has fallen at an average annual rate of 8.9 percent. The rapid substitution of BOF for OH furnaces has been the principal cause of this development; production by the BOF rose from 9 percent to 77 percent of the combined output of the two types of furnace. The BOF uses considerably less fuel per ton of steel than the older OH furnace. Unfortunately, no historical data on energy use in electric furnaces are available.

Heating and annealing furnaces use substantial amounts of fuel. We estimate, based on incomplete data, that they account for about one-fifth of the fuel use of the industry. During the 1963-76 period, fuel use in these furnaces per ton of raw steel rose (the same is true if shipments are used in the denominator). The rise was not great,

---

[f]Data assembled from reports of the AISI under VIECP.

[g]It should be noted that we are including coke as an energy source in blast furnaces, less any blast furnace gas (a by-product of coke combustion) that is used in other industry components.

Table 5-5.   Fuel Use in Components of the Steel Industry, 1963–76[a] (million Btu per ton)

	Blast Furnaces[b]		Open Hearth and Basic Oxygen Furnace (steel melting)		Heating and Annealing Furnaces	
	Coke per Ton of Pig Iron	Total Fuel Use per Ton of Pig Iron	Fuel Use per Ton of Raw Steel	Percent Produced by Basic Oxygen Furnace	Fuel Use per Ton of Raw Steel	Fuel Use per Ton of Products Shipped
	(1)	(2)	(3)	(4)	(5)	(6)
1963	18.0	18.5	3.1	8.7	4.0	5.7
1964	18.0	18.6	2.9	13.5	3.9	5.9
1965	17.6	18.2	2.7	19.4	4.1	5.8
1966	16.9	17.6	2.3	28.4	4.1	6.0
1967	16.4	17.1	2.0	37.0	4.3	6.6
1968	16.2	17.0	2.0	42.6	4.4	6.3
1969	17.0	17.8	1.7	49.7	4.4	6.6
1970	16.1	16.9	1.4	56.9	4.4	6.6
1971	16.5	17.4	1.1	64.3	4.8	6.4
1972	16.4	17.4	1.0	68.1	4.8	6.7
1973	16.2	17.3	1.0	67.7	4.6	7.0
1974	17.0	18.3	1.1	69.7	4.7	6.2
1975	15.2	16.5	0.9	76.4	5.1	6.3
1976	15.1	16.4	0.9	77.2	4.7	6.7

[a] Excludes electricity.
[b] Adjusted for change in coke inventory.

Source: Derived from data reported in American Iron and Steel Institute, *Annual Statistical Report*, various issues.

averaging only 1.2 percent or 1.3 percent per year, according to the output measure chosen.

Fuel Substitution

During the twenty-three years for which data are available in the AISI statistics, major changes were made in the fuel mix of the industry (Table 5—6). The use of coal declined in importance as did fuel oil to a lesser extent. Natural gas and purchased electricity, on the other hand, rose in importance.[h] The rise in the use of natural gas took place earlier than the rise in electricity; it was mainly accomplished by the early 1960s and represented a substitution of natural gas for coal and fuel oil in the generation of electricity and in "heating and annealing furnaces."

Increased electricity use is associated with two developments: the rise in the electric arc furnace and pollution control. The proportion of steel produced in EF rose from 12 percent in 1967 to 19 percent in 1976; we estimate that 45 percent of the rise in total electricity use in the steel industry came from the increased production of EF.[i]

In 1976, the use of natural gas in the industry fell sharply under the pressure of supply interruptions. There was also a rise in coal use—a reverse of the long-term trend shown in Table 5—6. Industry officials state that this will tend to *offset* gains in energy conservation in general because oil and natural gas are technically more efficient in many operations than coal. An example is the injection of oil in the blast furnace, which replaces coal with a higher Btu content per ton of iron produced.

Energy Use and Pollution Control

There are a number of aspects to environmental control in the steel industry that bear upon the problem of energy use. The most important of these are: (1) the steel industry is the largest polluter of the air and water in manufacturing; (2) the slow growth of the industry makes the cost of pollution abatement particularly burdensome because new facilities added each year are a very small fraction of existing capacity, and thus extensive and expensive retrofit is necessary to meet required standards; (3) nearly all energy use for pollution abatement is in the form of electricity; (4) nearly all the energy

---

[h]It should be recalled that we are measuring electricity at the heat rate (10,369 Btu per kwh in 1976, for example). A rise in purchased, and decline in self-generated, electricity, such as occurred in the steel industry, might otherwise be interpreted erroneously as a decline in total energy use.

[i]Using an electricity requirement of 550 kwh per ton, electric furnaces absorbed 8.3 billion kwh in 1967 (20 percent of the industry total) and 13.5 billion kwh in 1976 (25 percent).

Table 5—6.   Sources of Energy in the Steel Industry, 1953—76

				Coal for			
	Total Energy	Total	Purchased Electricity	All Fuels	Production of Coke[a]	Other Use	Fuel[b] Oil	Natural Gas
	(trillion Btu)			(Percent)				
1953	3459.2	100.0	7.6	92.4	70.2	5.9	9.6	6.7
1954	2794.7	100.0	7.1	92.9	70.1	5.9	8.9	7.9
1955	3477.2	100.0	7.8	92.2	69.8	5.1	9.1	8.1
1956	3424.7	100.0	8.1	91.9	69.0	5.2	9.2	8.6
1957	3489.3	100.0	7.7	92.3	70.7	5.4	7.8	8.3
1958	2648.0	100.0	6.7	93.3	67.0	7.0	8.4	10.9
1959	2750.3	100.0	6.7	93.3	66.2	6.4	8.7	11.9
1960	2914.5	100.0	6.9	93.1	65.7	6.5	8.1	12.8
1961	2748.6	100.0	7.5	92.5	63.0	6.8	7.8	15.0
1962	2780.9	100.0	8.0	92.0	62.3	6.5	7.1	16.2
1963	2923.0	100.0	8.4	91.6	61.6	6.1	7.5	16.4
1964	3332.9	100.0	8.1	91.9	63.5	5.5	7.0	15.9
1965	3476.9	100.0	8.4	91.6	63.4	5.5	6.4	16.3
1966	3469.1	100.0	9.0	91.0	64.3	5.5	5.8	15.4
1967	3346.3	100.0	9.5	90.5	63.5	5.3	5.2	16.5
1968	3405.7	100.0	10.2	89.8	61.6	5.1	5.2	17.8
1969	3608.2	100.0	10.6	89.4	62.2	4.2	4.8	18.2
1970	3545.6	100.0	11.2	88.8	62.9	3.7	4.9	17.3
1971	3246.3	100.0	12.2	87.8	60.1	3.7	4.8	19.2
1972	3413.0	100.0	12.1	87.9	60.0	3.3	5.4	19.3
1973	3775.2	100.0	12.7	87.3	60.5	3.3	5.9	17.6
1974	3743.2	100.0	12.9	87.1	59.3	2.7	6.6	18.5
1975	3306.2	100.0	12.7	87.3	57.7	2.4	6.1	21.1
1976	3329.4	100.0	13.8	86.2	58.5	2.4	6.8	18.5

[a] Adjusted for change in coke inventory.
[b] Includes small amounts of liquefied petroleum gas.
Source: Derived from data reported in American Iron and Steel Institute, *Annual Statistical Report*, various issues.

use for environmental control has occurred since 1967, and most of it since 1971.

There are no historical data on energy use for pollution control by the steel industry to our knowledge. At least three studies on the problem have been undertaken but all were directed to the question "how much must energy use rise in the steel industry in order to comply with pollution regulations?" They all deal with a hypothetical situation and do not attempt to measure annual energy usage for pollution control in the industry. One of these studies indicates, however, that nearly all the energy use for pollution abatement is in the form of electricity [7].

On the basis of the limited information available, we have prepared an estimate of energy use for pollution control (all of which is assumed to be in the form of electricity)—this is summarized below.

	1967	1976	Change (2) − (1)
	(1)	(2)	(3)
(a) Total energy use (trillion Btu)	3,346	3,329	−17
(b) Product shipments (million tons)	83.90	89.45	+6.55
(c) Energy use per ton (million Btu)	39.88	37.22	−2.66
(d) Total electricity use (billion kwh)[j]	42.5	54.3	+11.8
(e) Electricity use in EF (billion kwh)	8.3	13.5	+5.2
(f) Other electricity use (billion kwh)	34.2	40.8	+6.6
(g) Other electricity use (trillion Btu)	356	423	+67
(h) Row (g) as percent of row (a)	10.6	12.7	+2.1
(i) Energy use per ton, holding "other electricity use" at 1967 level	39.88	36.46	−3.42

We conclude that energy use for pollution control accounted for *at most* 2.1 percent of total energy use in 1976 [8]. Furthermore, energy use per ton of shipments would have fallen *at most* by 1.0 percent per year from 1967 to 1976 instead of the actual decline of 0.8 percent per year if pollution control had not risen above the 1967 level.

Determinants of Energy Conservation

We have found that no progress in energy conservation was made in the steel industry since 1974; in fact, a worsening probably occurred. The principal cause of this development was the sharp decline in output during the 1975 recession, which was only partially re-

[j]Purchased and self-generated electricity.

coupled in 1976. Other negative factors in recent years have been energy use for environmental control and interruptions in the supply of natural gas, causing inefficient fuel substitution.

On the positive side, a number of influences that can aid energy conservation can be noted. One of these is the rise in continuous casting. This method uses less energy than the usual method, owing to the reduced need for reheating. But more important is the fact that it reduces scrappage, thus raising the yield of a ton of steel with attendant energy savings [9]. The AISI has kindly made the following unpublished data on continuous casting available to us.

	Continuous Casting as Percent of Raw Steel Production
1970	3.8
1972	5.8
1974	8.1
1976	10.6
1977 (11 months)	11.8

These figures indicate a substantial source of energy saving.

Another effective means of conserving energy is a rise in scrap use because it does not entail the heavy energy use that occurs in ore preparation and, above all, in the blast furnace; such a rise has taken place over the 1963–76 period (9.8 million tons). Scrap accounts for nearly all the material inputs of the EF, so even though the EF uses more energy to melt steel than either the OH or the BOF, it has served as an energy saver on balance by consuming more scrap over the period (14.6 million tons). The BOF has a lower technical limitation on the proportion of its charge made up of cold scrap than the OH. As the OH has declined in importance, the EF has absorbed the excess scrap formerly consumed in the OH and not taken up by the BOF (4 million tons).

The last development we cite is the shift away from the OH. Not only does the OH use more energy than the BOF, but it also has more severe pollution problems, which require additional energy use to abate. The recent closings of OH installations have caused serious local employment problems, but they have helped to promote energy conservation in the industry.

Future trends in energy conservation in the industry will depend largely on the rate of investment in new facilities. The technology necessary for making substantial gains in energy conservation is well known. But output of steel has been depressed since 1974, and a reasonably high rate of growth of output is necessary to make investment profitable in facilities that incorporate this new technology.

Conclusions

Energy use per unit of output rose from 1973 to 1976. This result is found regardless of the output measure employed. This worsening resulted from three main influences. The most important is the sharp drop in capacity utilization, lowering efficiency of energy use. A second was interruptions in supplies of natural gas, which caused production stoppages and inefficiencies brought about by the necessity for rapid fuel substitution. Pollution control, mainly in the form of electricity, also raised total energy use.

These three negative factors more than offset the savings won through a rise in continuous casting, sharply increased use of scrap, and a continuing decline in open-hearth steelmaking.

PRIMARY ALUMINUM (SIC 3334)

This industry is the most energy intensive of the top twenty energy users in manufacturing (see Table 2–5). It is one of the fastest growing of the basic materials industries, and production is carried on in a small number of plants (thirty-one in 1972) by a small number of companies (ten in 1972).

Boundaries of the Industry

The term "aluminum industry" is often used to refer to a series of activities ranging from the mining of bauxite to the production of cans, pipe, siding, and other finished or semifinished products. In this book, we concentrate on the basic product: producing metal from alumina and forming it into ingots or billets, which is termed "primary aluminum" in the SIC.

The comparatively small amount of bauxite that is produced in the United States is classified in the mining sector; production of alumina from (mainly imported) bauxite is classified in industrial inorganic chemicals, n.e.c. (SIC 2819), which is discussed in Chapter 4. Further fabrication of the ingots and billets and the processing of aluminum scrap are classified in other manufacturing industries.

The primary aluminum industry is quite highly specialized, producing ingots and billets almost exclusively (97 percent of shipments in 1972). About one-fourth of all aluminum ingots and billets are produced in other industries, however; most of this other production is from scrap.

Production Processes

The Hall-Heroult process for smelting alumina into metal has been the only process in commercial use in this century [10]. A new plant that will use a chemical reduction process has been under construction in recent years but did not contribute to output during the period under review.

Forming the molten metal into ingots and billets, often referred to as "holding, casting, and melting," is characterized by a number of techniques that are largely determined by the type of furnace used. Most aluminum is alloyed in these furnaces and then formed into ingots and allowed to cool for shipment to fabricating plants. For some products, it is possible to omit the formation of ingots and the necessity for cooling and reheating by transporting the molten metal directly from the smelter to continuous rod and sheet mills or casting operations. In the past, transportation costs dictated the location of smelters near inexpensive energy supplies and fabricating plants near market areas. High energy prices will probably change this pattern for many new plants. The savings in energy through continuous operations will more than offset the added transportation cost, leading to the location of fabricating facilities near the smelter.

The electrolytic smelting process is by far the largest user of energy in the entire sequence from mining to fabrication, and thus developments in that process tend to dominate total energy use. Data furnished to the BOM by the Aluminum Association show that 67 percent of all energy used in the aluminum industry, from bauxite mining through fabrication, was in hot metal production [11]. Holding, casting, and melting, the other part of primary aluminum, accounted for only 6 percent of the total.

Measurement of Output and Energy Use

Because of the standardized, homogeneous nature of the products and the high degree of specialization of this industry, there is practically no difference in behavior over time between a weighted index of production and a simple quantity measure. We have therefore decided to use production of primary aluminum in pounds as our measure of output.[k] Physical production and capacity are published annually by the Aluminum Association.

In Chapter 2, there was a brief discussion of the appropriate measurement of electricity. There we maintained that it was desirable to "go back to the powerhouse" by converting kwh of electricity pur-

---

[k]This treatment avoids the difficulty caused by the census output index for the 1967-72 period, which apparently understates the rise in production in this industry.

chased by the "heat rate," that is, the average number of Btu used by electric utilities to produce a kwh. This average is, however, for fossil-fuel utilities, while a substantial fraction of the electricity used in all parts of the aluminum industry is generated by hydropower [12].

We have decided to convert purchased electricity generated by hydropower at the fossil-fuel rate. While this conflicts with industry usage, in which hydropower-generated electricity is converted by the thermal content (3,412 or 3,413 Btu per kwh), we believe our treatment to be preferable for two reasons: (1) The possibilities for expansion of hydropower are limited in the United States, and thus an additional kwh used is likely to be generated by fossil fuels (or possibly nuclear energy). Our method thus converts hydropower-generated electricity at its replacement rate. (2) When all purchased electricity is converted at the same rate, variations from period to period in the mix of power sources used to generate that electricity do not cause the apparent energy use of the industry to fluctuate.

Appropriate treatment of electricity is very important for the primary aluminum industry because of its heavy reliance on this source. From the data in Table 5−7, we estimate that about 93 percent of all energy consumed in 1976 came from electricity (86 percent from purchased and 7 percent from self-generated electricity).

Growth and Fluctuation

During the years from 1947 to 1967 the industry grew rapidly at an average rate of 9.4 percent per year. The number of firms expanded from three in 1947, to ten in 1967, and then to twelve in 1976. Since 1967, output growth has slowed somewhat, averaging 6.0 percent to 1974.

These rapid average rates of growth mask the very wide swings in demand to which the industry has been subjected. As shown in Table 5−7, production has ranged from a high of 104 percent of stated capacity (at year-end) to a low of 71 percent during a recession. The energy use of the industry is strongly affected by these fluctuations in product demand.

The Trend in Energy Conservation, 1947−76

From 1947 to 1967, Btu per pound of primary aluminum fell 34 percent, or 2.1 percent per year (Table 5−7). But from 1967 to 1976, the rate of decline was only 0.9 percent per year. Part of the explanation for this slowdown may be seen by comparing gross energy per pound with kwh per pound (including self-generated electricity). From 1947 to 1967, the more rapid decline of the gross

Table 5-7. Energy and Output in Primary Aluminum

	Production (million pounds)	Capacity at Year-end (million pounds)	Capacity Utilization (1) ÷ (2) (percent)	Gross Energy (trillion Btu)	Gross Energy per Pound of Production (4) ÷ (1) (1,000 Btu per pound)
	(1)	(2)	(3)	(4)	(5)
1947	1,143.5	NA	NA	156.5	136.9
1948	1,246.9	1,283.0	97.2	NA	NA
1949	1,206.9	1,308.0	92.3	NA	NA
1950	1,437.2	1,502.5	95.7	NA	NA
1951	1,673.8	1,601.5	104.5	NA	NA
1952	1,874.7	2,311.4	81.1	NA	NA
1953	2,504.0	2,671.4	93.7	NA	NA
1954	2,921.1	2,826.4	103.4	321.4	110.0
1955	3,131.4	3,269.4	95.8	NA	NA
1956	3,357.9	3,551.0	94.6	NA	NA
1957	3,295.4	3,678.0	89.6	NA	NA
1958	3,131.1	4,388.5	71.3	303.3	96.9
1959	3,908.2	4,805.5	81.3	NA	NA
1960	4,029.0	4,937.5	81.6	NA	NA
1961	3,807.4	4,967.5	76.6	NA	NA
1962	4,235.9	4,977.5	85.1	424.6	100.2
1963	4,625.1	5,021.5	92.1	NA	NA
1964	5,105.5	5,198.2	98.2	NA	NA
1965	5,509.0	5,516.6	99.9	NA	NA
1966	5,936.7	6,330.6	93.8	NA	NA
1967	6,538.5	6,642.0	98.4	588.9	90.1
1968	6,510.1	7,372.0	88.3	NA	NA
1969	7,586.1	7,768.6	97.6	NA	NA
1970	7,952.3	8,430.0	94.3	NA	NA
1971	7,850.4	9,332.0	84.1	595.7[a]	75.9[a]
1972	8,244.8	9,542.0	86.4	NA	NA
1973	9,058.2	9,786.0	92.6	NA	NA
1974	9,806.9	9,832.0	99.7	864.3	88.1
1975	7,758.3	10,043.0	77.3	679.7	87.6
1976	8,502.8	10,385.8	81.9	705.7	83.0

(continued from above)

	Electricity Use (million kwh)			*Kilowatt-hour per pound*
	Purchased	*Self-generated*	*Total*	$(8) \div (1)$
	(6)	(7)	(8)	(9)
1947	9,331	1,070	10,401	9.10
1948	NA	NA	NA	NA
1949	NA	NA	NA	NA
1950	NA	NA	NA	NA
1951	NA	NA	NA	NA
1952	NA	NA	NA	NA
1953	NA	NA	NA	NA
1954	17,239	9,044	26,283	9.00
1955	NA	NA	NA	NA
1956	NA	NA	NA	NA
1957	NA	NA	NA	NA
1958	16,170	8,968	25,138	8.03
1959	NA	NA	NA	NA
1960	NA	NA	NA	NA
1961	NA	NA	NA	NA
1962	26,883	9,387	36,270	8.56
1963	NA	NA	NA	NA
1964	NA	NA	NA	NA
1965	NA	NA	NA	NA
1966	NA	NA	NA	NA
1967	41,957	11,648	53,605	8.20
1968	NA	NA	NA	NA
1969	NA	NA	NA	NA
1970	NA	NA	NA	NA
1971	42,712[a]	10,976[a]	53,688[a]	6.84[a]
1972	49,362	11,517	60,880	7.38
1973	59,308	11,406	70,715	7.81
1974	68,699	8,718	77,417	7.89
1975	55,624	4,968	60,592	7.81
1976	58,777	4,785	63,562	7.48

(Table 5-7. continued overleaf)

Table 5–7. continued

	Production (million pounds)	Capacity at Year-end (million pounds)	Capacity Utilization (1) ÷ (2) (percent)	Gross Energy (trillion Btu)	Gross Energy per Pound of Production (4) ÷ (1) (1,000 Btu per pound)
	(1)	(2)	(3)	(4)	(5)
	Average Annual Rate of Change (percent)				
1947–67	9.1	9.0[b]	—	6.9	-2.1
1967–76	3.0	5.1	—	2.0	-0.9
(1967–74)	(6.0)	(5.8)	—	(5.6)	(-0.3)
(1974–76)	(-6.9)	(2.8)	—	(-9.6)	(-2.9)

[a] The 1971 energy figures for this industry, derived from data in the 1971 *Annual Survey of Manufactures*, are much too low.
[b] Source: The Aluminum Association, *Aluminum Statistical Review 1975; Census of Manufactures; Annual Survey of Manufactures*.

Table 5–7. continued

	Electricity Use (million kwh)			Kilowatt-hour per pound (8) ÷ (1)
(continued from above)	Purchased	Self-generated	Total	
	(6)	(7)	(8)	(9)
	Average Annual Rate of Change (percent)			
1947–67	7.8	12.7	8.5	-0.5
1967–76	3.8	-9.4	1.9	-1.0
(1967–74)	(7.3)	(-4.0)	(5.4)	(-0.5)
(1974–76)	(-7.5)	(-25.9)	(-9.4)	(-2.6)

energy ratio than of the kwh ratio resulted from improvements in the heat rate, both by electric utilities and within the industry. Over the 1967−76 period, the heat rate did not change appreciably, and the kwh ratio showed an accelerated rate of decline when compared with the preceding twenty years. The source of the more rapid savings in the earlier (1947−67) period was increased efficiency of electricity generation, both within and outside the industry. This factor is not important in the later (1967−76) period, so that we use kwh per pound as our primary measure because of the availability of data for more single years.

When the 1967−76 period is divided into two parts, 1967−74 and 1974−76, roughly pre- and post-embargo, a sharp break is evident. During the first six years, kwh per pound fell at the long-term pace, 0.5 percent per year. Thereafter, the rate of energy saving accelerated sharply to 2.6 percent per year from 1974 to 1976. This pattern is also evident in the ratio of gross energy to constant dollar value added or to pounds of production.

An Alternative Measure of Energy Saving

The Aluminum Association has submitted data on energy use to the federal government under VIECP [13]. The statistics cover the 1972−76 period and provide an alternative measure of energy use. From 1972 to the second half of 1976, the Aluminum Association reported a saving of more than 7 percent. However, most of these savings were in fabrication. A computation from these reports of energy use in the "hot metal" and "holding, casting, and melting" segments, which make up the primary aluminum industry, yields the following results: the primary aluminum industry showed a slight worsening from 1972 to 1974 in Btu per pound, but a sharp acceleration thereafter from 1974 to 1976. These results are in substantial agreement with our findings despite differences in concept, coverage, and conversion factors.

Sources of Energy Savings

The statistical results indicate a significant response to the energy problems of the post-embargo period. It is of considerable interest to seek the ways in which this apparent energy saving was obtained. Discussions with industry officials, reading of the literature, and inspection of the data suggest three sources:

(1) **New Plant and Equipment.** The rate of expansion in the industry slowed substantially in the 1967−74 period compared with the preceding twenty years. In the 1974−76 period, the rate

of expansion was further halved as a result of the recession and the accompanying abrupt decline in capacity utilization. While some improvement in energy use can be attributed to the 5.6 percent net expansion in capacity from 1974 to 1976 and to rebuilding of existing facilities, only a small part of the total improvement could have come from this source.

(2) **Cyclical Response.** The decline in capacity utilization by more than 22 percentage points in 1975 was followed by only a partial recovery (4.6 percentage points) in 1976. The industry responded to the decline in demand by shutting down many smelters in whole or part. An inspection of the data in Table 5−7 suggests that kwh per pound, our measure of energy conservation, does not rise when capacity utilization falls, but it either falls or remains unchanged. This is quite different from the pattern found in other industries, such as steel, where more energy per unit of output is used in periods of low-capacity utilization.

We draw two inferences from this pattern. The first is that energy conservation in the industry can best be measured between years with high (or at least similar) capacity utilization rates. The second is that the industry adjusts to reduced demand by closing down the least efficient (presumably the oldest) operations, either entire plants or components (potlines); this causes the kwh-output ratio to fall. Such a response to a drop in demand is more easily accomplished in an industry with only a few firms who control large fractions of total output and can, therefore, exercise a wider choice of how a given volume is to be produced.

Several primary aluminum plants shut down completely during 1975, and others reduced their output substantially. It seems reasonable to assume that the least efficient operations were closed, thus contributing substantially to the drop in the kwh per pound of aluminum ratio.

(3) **Recycling.** Use of scrap reduces energy use substantially. An earlier study put the energy "cost" of producing a pound of molten metal from scrap at only 6,000 Btu per pound [14]; this may be roughly compared with the 83,000 Btu per pound figure shown in Table 5−7. From 1974 to 1976, primary producers of aluminum increased their consumption of scrap very sharply [15]. This undoubtedly contributed to the accelerated energy savings of those years.

In summary, we conclude that scrap utilization and the closing of the least efficient facilities were the main causes of the acceleration of energy savings during the 1974—76 period. While scrap utilization can be maintained and expanded, the cyclical effect is temporary.

A Negative Factor—Pollution Control

Smelting alumina into primary aluminum creates serious problems of air pollution in the form of fluorides and particulate matter. A substantial amount of energy is used in controlling air emissions. Historical data on this energy use are incomplete at this writing. From data in a study published by the Department of Commerce [16], we estimate that 1.6 percent of energy use in primary aluminum in 1972 was for pollution control and that this had risen to 2.8 percent by 1976. This served to slow improvements in energy savings in the industry and will continue to do so until the progressively more strict timetable of pollution abatement requirements for the industry have been fully complied with.

Prospects for Future Conservation

Little can be done to reduce the energy use in reducing alumina to aluminum ingot or billet in an existing smelting facility. Savings can be achieved by using more scrap and by the introduction of continuous casting (where plant layout and product mix permit). But once a potline is installed, it cannot be altered without extensive rebuilding, which is seldom economical.

New smelting facilities are a different matter. Proven methods exist for producing aluminum with about 20 percent less energy than the present average. The chemical reduction process mentioned earlier also promises substantial savings beyond the 20 percent mentioned. But this can be brought about only with capacity expansion and turnover. Many existing facilities, which incorporated the technology that was most up-to-date at the time of their construction, use much more energy than the current state of the art.

A continuation of the slow growth of recent years will make it very difficult for the primary aluminum industry to continue to make important energy savings. Only expansion and replacement of existing facilities will permit substantial improvement in the future.

Conclusions

Energy conservation in primary aluminum accelerated sharply in the 1974—76 period compared with the preceding seven years. Two important determinants of this improvement appear to be the shutting down of the least efficient production facilities and a substantial rise in recycling of scrap aluminum. A rise in energy use for air pollution control, an important element in this industry, offset part of the savings.

NOTES TO CHAPTER 5

1. Council on Wage and Price Stability, *Prices and Capacity Expansion in the Cement Industry*, March 1977. Mimeographed.

2. Bureau of Mines, *Minerals Yearbook*, vol. 1, "Cement."

3. Battelle-Columbus Laboratories, *Energy Use Patterns in Metallurgical and Nonmetallic Mineral Processing*, Phase 4–Energy Data and Flowsheets, High-Priority Commodities, NTIS, 1975, p. 25.

4. Portland Cement Association, *The Potential for Energy Conservation in the Cement Industry*, Conservation Paper No. 26, U.S. Department of Commerce, June 1975. Energy use by age of plant and process are presented on p. 18.

5. *Voluntary Industrial Energy Conservation, Progress Reports*, U.S. Department of Commerce and Federal Energy Administration.

6. Estimate derived from tables in Battelle-Columbus Laboratories, *Evaluation of the Theoretical Potential for Energy Conservation in Seven Basic Industries*, NTIS, 1975, pp. 81–92.

7. Resource Planning Associates, *Energy Requirements for Environmental Control in the Iron and Steel Industry*, NTIS, January 1976, p. 19, indicates that 96 percent of the *cost* of energy use for pollution control would be spent on electricity.

8. This estimate agrees closely with an estimate in Battelle-Columbus Laboratories, *Potential for Energy Conservation in the Steel Industry*, May 30, 1975.

9. The average yield in a slab mill is estimated at 85 percent, in continuous casting 96 percent. Battelle-Columbus Laboratories, *Energy Use Patterns in Metallurgical and Nonmetallic Mineral Processing*. The saving in energy per ton of slab steel by the continuous process is substantial, pp. 66–71.

10. For a description of the process and the various plant configurations see Richard J. Herbst and Robert B. Grant, *Energy Requirements for Air Pollution Control in the Primary Aluminum Industry*, U.S. Department of Commerce, 1977.

11. "Aluminum," preprint from Bureau of Mines, *Minerals Yearbook*, Vol. 1, 1974, p. 23.

12. About 38 percent in 1972, according to the Aluminum Association. See "Aluminum" preprint from 1974 *Minerals Yearbook*.

13. *Voluntary Industrial Energy Conservation, Progress Reports*.

14. M.F. Elliott-Jones, "Aluminum" in J.G. Myers *Energy Consumption in Manufacturing*.

15. The Aluminum Association, *Aluminum Statistical Review*, 1976, pp. 38–9.

16. Herbst and Grant, *Energy Requirements for Air Pollution Control in the Primary Aluminum Industry*.

Appendixes

Energy Conversion Factors

In this appendix we discuss the energy conversion factors used in this book to convert reported consumption of fuels and electric energy to the standard energy measure, British thermal units (Btu). First, we shall discuss the conversion factors that we applied to energy data from the Census Bureau. These data are the basis for Chapter 2 and for the reports on paper and paperboard, organic and inorganic chemicals, and aluminum. Second, we shall discuss the conversion factors we used for energy data published by the Bureau of Mines (BOM) and various industry trade associations. These data are particularly important in the reports on petroleum refining, blast furnaces and steel mills, and hydraulic cement. The final section presents conversion factors used for purchased electric energy.

CENSUS BUREAU DATA

For Chapter 2 and for five of our eight industries, our analyses are based on Census Bureau data. These data were collected for 1947, 1954, 1958, 1962, 1967, and 1971 in conjunction with the periodic *Census of Manufactures*. Annual collection of detailed energy data was begun in 1974; the most recent figures available are for 1976.

Our primary source for energy conversion factors is *Energy Consumption in Manufacturing*.[1] Our general energy conversion factors for coal, coke and breeze, distillate and residual fuel oil, and natural

1. See reference, p. 6n.

gas (shown in Table A–1) are taken from pages 84 and 85 of that report.

It should be noted that the bituminous coal conversion factor is applied to the census category "bituminous coal, lignite and anthracite" and that the coke conversion factor is applied to the category "coke and breeze." In general, these approximations should cause only very slight errors, but the error might be important in individual industries.

Distillate and residual fuel oil are converted individually* for years 1962 and forward. Prior to 1962, the Census Bureau did not publish fuel oil by type, but only as a total. For this reason, total fuel oil prior to 1962 is converted by using the 1962 weighted average of residual and distillate conversion factors for each industry.

"Other fuels" and "fuels, not specified by kind (n.s.k.)," present a special problem. For both these categories, only dollar expenditures are published. Other fuels are fuels other than those specifically designated on the energy data collection forms. (Fuels specifically designated are bituminous coal, lignite, anthracite, coke, breeze, natural gas, residual fuel oil, and distillate fuel oil.) This category includes gasoline, propane, butane, steam, and wood. Fuels, n.s.k., are fuel expenditures that are not reported in detail or that are estimated by the Census Bureau for nonreporting small firms.

Published totals for other fuels and fuels, n.s.k., have fluctuated substantially both as proportions of total fuel expenditures and in relation to each other. In 1962, for example, other fuels were 10 percent of all manufacturing fuels purchased but only 5.5 percent in 1967. These fluctuations have resulted from inconsistent treatment by the Census Bureau over the years, particularly before 1971.

When a firm fails to report details of fuel use or reports details of fuel use ambiguously, the Census Bureau has the option of following up and getting more detailed information, calling the total fuels, n.s.k., distributing the unknown sum proportionately over the fuel categories, or calling the total other fuels. The procedures followed by the Census Bureau have apparently varied widely from census to census; as a consequence, the classification of expenditures into one category or another frequently has depended more on census methods than on industry practices.

The Census Bureau converts fuels, n.s.k., based on the average price paid for fuels in the previous survey and carried forward using the Bureau of Labor Statistics' Wholesale Price Indexes for fuels and

---

*This practice was not followed in *Energy Consumption in Manufacturing* and causes slight differences between that publication and this book.

Table A–1. Fuel Conversion Factors

Fuel	Use	Conversion Factor	Units
Coal	General	25.8	Million Btu per short ton
	Coal for Coke (SIC 3312)	26.0	" " " " "
	Petroleum Refining (SIC 2911)	26.0–22.8	" " " " "
Coke	General	26.0	" " " " "
Fuel oil (unspecified)	General	5.8	Million Btu per barrel
	Petroleum Refining (SIC 2911)	6.287	" " "
Fuel oil, distillate	General	5.825	" " "
Fuel oil, residual	General	6.287	" " "
Natural gas	General	1035	Btu per cubic ft.
	Petroleum Refining (SIC 2911)	1020–1050	" " " "
Acid sludge	Petroleum Refining (SIC 2911)	4.5	Million Btu per barrel
Refining gas	Petroleum Refining (SIC 2911)	940–1500	Btu per cubic ft.
Liquefied petroleum gases	Petroleum Refining (SIC 2911)	4.0–4.3	Million Btu per barrel
Petroleum coke	Petroleum Refining (SIC 2911)	30.12	Million Btu per short ton
Coke oven gas	Blast furnaces and Steel Mills (SIC 3312)	550	Btu per cubic ft.
Tar and pitch	" " " " " "	6.0	Million Btu per barrel
Blast furnace gas	" " " " "	95	Btu per cubic ft.

Note: General use conversion factors are used for all Census Bureau energy data. Special industry conversion factors are applied only to special industry energy data sources.

Source: *Energy Consumption in Manufacturing*; Bureau of Mines; American Iron and Steel Institute.

power. These conversion factors are calculated separately for each two-digit industry. Other fuels are converted by a special factor estimated by the Census Bureau. In 1975, the Census Bureau's specific conversion factors were $2.10 per million Btu for other fuels and an average of about $1.30 per million Btu for fuels, n.s.k. This method, which gives considerably different conversion factors to other fuels and fuels, n.s.k., means that the energy totals include a fluctuation caused by the inconsistency of census procedures.

The procedure we have followed is to add the two together and to assume that they have the same average Btu content per dollar in a given census year as the specified fuels. This is done separately at the four-digit and two-digit levels. For all manufacturing, however, there is an aggregation problem. Our procedure was to convert other fuels and fuels, n.s.k., separately for the five two-digit industries we are studying and for all other two-digit industries. The sum of these two is the total for all manufacturing industries.

DATA SUPPRESSION

The Census Bureau collects data under a statutory requirement of confidentiality. Wherever the publication of data might permit information about a specific company or plant to be determined by another company, that statistic is suppressed.

For example, if only two companies use coal for heat or power in a specific four-digit industry, the coal data for the industry is suppressed. At the same time, because industry totals are shown for all fuels consumed, the data for another category (usually other fuels or fuels, n.s.k.) must also be suppressed; otherwise, the coal data could be obtained by subtraction. Thus the total for two categories is known but not the breakdown between the two. In addition, the industry group totals (at the two-digit level) are published. This requires that coal data be suppressed for some other industry in the industry group. Again, the total for the two industries is known but not the breakdown within the two.

The combination of the two pieces of information can often place reasonable limits on each datum. For example, if the purchases of coal and fuels, n.s.k., for a single industry total $42 million and total coal expenditures for that industry and another are $4 million, then coal must be between $0 and $4 million and fuels, n.s.k., between $38 and $42 million. In some cases, where dollar totals for suppressed data were known to be small percentages of total fuel purchases, they were included in fuels, n.s.k., or other fuels.

The only industry for which we did not estimate suppressed data was the primary aluminum industry. Here we simply accepted Census Bureau conversion factors for all fuels and their published total of kwh equivalents for all fuels.

SPECIAL INDUSTRY DATA

In three industries, petroleum refining, blast furnaces and steel mills, and hydraulic cement, special data sources were used as the primary basis for analysis. In these cases, it was necessary to adopt special conversion factors for fuels. Table A−1 shows the conversion factors that varied from our general energy conversion factors.

The BOM, our source for the conversion factors used in the petroleum refining industry, has changed these factors over time to reflect variations in average heat content for each energy source. We have accepted this set of energy conversion factors for the industry.

Additional energy data were obtained from trade association reports under VIECP. In analyzing these reports, we have generally accepted the conversion factors developed by the reporting firms, the only exception being electrical energy. Electrical energy has uniformly been converted using heat rates.

ELECTRICAL ENERGY

Kwh of purchased electrical energy are converted to Btu by the fossil-fuel heat rate for the year. This represents the average fuel use in fossil-fuel electric utility power plants for each kwh of electricity generated. This conversion factor, available from the Edison Electric Institute, is based on data published by the Federal Power Commission. Table A−2 shows heat rates for 1947 to 1976. For a discussion of this measure, including its justification and interpretation, see Chapter 2.

(Table A−2 overleaf)

Table A–2.   Heat Rate, Electricity Conversion Factor

Year	Btu per kwh	Year	Btu per kwh
1947	15,600	1962	10,493
1948	15,738	1963	10,438
1949	15,033	1964	10,407
1950	14,030	1965	10,384
1951	13,641	1966	10,399
1952	13,361	1967	10,396
1953	12,889	1968	10,371
1954	12,180	1969	10,457
1955	11,669	1970	10,508
1956	11,456	1971	10,536
1957	11,365	1972	10,479
1958	11,090	1973	10,429
1959	10,879	1974	10,481
1960	10,701	1975	10,383
1961	10,552	1976	10,369

Source: Edison Electric Institute.

※ *Appendix B*

Output Indexes

In this appendix, we discuss the construction of our combined Census Bureau-Federal Reserve Board (FRB) output indexes and some of the technical problems of the indexes.

METHOD OF CONSTRUCTION

In 1977, the Census Bureau published "Indexes of Production," Volume 4 of the *Census of Manufactures, 1972*. The indexes in this volume provide the general base for our estimates of constant dollar value added for the period 1967–76. The published indexes show the relative change in production between census years, in this case 1972 compared with 1967. The particular indexes we have used in this report are cross-year value-added weighted indexes [1].

To obtain output measures for years other than census years, we used the annual FRB indexes of production. These are interpolated for 1968 to 1971 and extrapolated from 1973 forward. Having obtained relative indexes (with 1967 set arbitrarily at 100), we then converted the data to 1974 dollars. This was done by multiplying the entire series by the ratio of the 1974 index number to 1974 value added as published in the Census Bureau's *Annual Survey of Manufactures*.

To illustrate the method of construction, we show in Table B–1 a hypothetical calculation. The census index for our hypothetical industry in 1972 is 115.5 (1967 = 100). This is shown in row 1. The corresponding FRB index for this example is shown in row 2. Row 3 shows the ratios used to interpolate the census index and to extrapo-

Table B-1. Illustration of Method of Construction for Constant Dollar Output

	1967	1968	1969	1970	1971	1972	1973	1974	1975	1976
Census index	100.0					115.5				
FRB index	100.0	100.0	102.0	104.0	105.0	110.0	115.0	120.0	125.0	130.0
Ratio	1.000	1.010	1.020	1.030	1.040	1.050	1.050	1.050	1.050	1.050
Census-FRB index	100.00	101.00	104.04	107.12	109.20	115.50	120.75	126.00	131.25	136.50
Value added (million dollars)								252.0		
Output (million 1974 dollars)	200.0	202.0	208.1	214.2	218.4	231.0	241.5	252.0	262.5	273.0

late it. From 1967 and 1972, we used a straight line to interpolate between the end years. We assumed that the FRB index correctly indicates relative change for years after 1972, and we simply adjusted the level of the indexes to the 1972 census level. Row 4 shows the combined Census Bureau-FRB index with 1967 set at 100. Row 5 shows the 1974 value added for this hypothetical industry. Row 6 shows the Census Bureau-FRB index expressed in 1974 dollars. To create this series, row 5 has been multiplied by $2 million, the ratio of 1974 value added to the index number.

TECHNICAL PROBLEMS

The gross output of manufacturing industries is produced from inputs of labor, capital, and materials (including energy). Because the materials consumed in production are themselves the products of other industries, they must be subtracted from industry output to avoid double counting; industry output less materials consumed is called value added.

Neither the FRB indexes nor the census indexes are true value-added indexes. They use fixed value-added weights—that is, they are based on the levels of value added—but changes in the ratio of value added to gross output from year to year are not taken into account.

This causes errors in the production indexes whenever the degree of integration within an industry changes. For example, if a pulp mill is added to a nonintegrated papermill to create an integrated pulp and papermill, the added output of the pulp mill is not included in the census output index.

Organic Chemicals

A variation in this problem occurred with unusual force in the organic chemical industry between 1967 and 1972. During this period, chemical companies frequently found that as a result of capacity imbalances it was cheaper to purchase intermediate chemicals than to produce them. As a result, total pounds sold of chemicals rose far more rapidly than did pounds produced. In the largest category, "miscellaneous chemicals, acyclic," sales rose from 25.2 billion pounds to 44.0 billion pounds, an increase of 74 percent, while production rose from 58.2 billion pounds to 88.1 billion pounds, or 51 percent.

As a consequence, the census output index from 1967 to 1972 was considerably overstated. For our analysis in Chapter 4, it was necessary to adjust the census output index to produce an accurate relationship between energy and output. To do this, we reduced the sales of miscellaneous chemicals, acyclic, to correspond to the change in production. As a result, the index for 1972 is revised from 175.6[a] to 163.1.

This adjustment was not included in Chapter 2, our summary chapter. To change one industry without changing others carries a strong risk of inconsistency of treatment. Since the error in the organic chemical industry has only a very modest effect on total manufacturing, we have allowed the original census figures to stand. (The correction for all manufacturing would be to decrease 1972 production—relative to 1967—by 0.1 percent).

NOTE TO APPENDIX B

1. For an explanation of these terms and the method for calculation of the census indexes, see *Census of Manufactures, 1972*, Volume 4, "Indexes of Production," U.S.G.P.O. 1977.

---

[a]This number is already adjusted downward from the published census figure (176.8) to include urea production. See Chapter 4 for further details on changes in census classifications in chemical industries.

Survey of Twenty Large-Volume Petrochemicals

Data on production of twenty of the largest volume petrochemicals were obtained from the U.S. International Trade Commission annual publication, *Synthetic Organic Chemicals*. Selected years from 1967 to 1976 are presented in Table C-1. A wealth of technical information on production costs is available in industry journals, chemical handbooks, and other sources. The Stanford Research International series of publications on process economics is perhaps the single most valuable source.

These sources were combed for energy use by chemical process in a study performed in 1973 and published in *Energy Consumption in Manufacturing*. Estimates were made of average practice in 1954 and 1967 and projections made for 1980. Table C-2 shows estimated energy use for chemicals for various years, obtained by straight-line interpolation of the trend between 1967 and 1980.

Table C-1.   Production of Twenty Selected Large-Volume Petrochemicals, Selected Years, 1967-76 *(million pounds)*

	1967	1971	1974	1975	1976
Acetic acid	1,560	1,956	2,584	2,197	2,463
Acetic anhydride	1,556	1,513	1,633	1,458	1,506
Acetone	1,284	1,538	1,980	1,640	1,869
Acrylonitrile	671	979	1,412	1,214	1,518
Adipic acid	971	1,306	1,478	1,343	1,281
Carbon disulfide	694	753	782	479	508
Ethyl alcohol	1,918	1,630	1,618	1,429	1,496
Ethyl chloride	618	620	662	575	669
Ethylene glycol	1,989	3,070	3,341	3,809	3,335
Ethylene oxide	2,308	3,598	3,892	4,467	4,184
Formaldehyde	3,707	4,522	5,764	4,558	5,449
Hexamethylene diamine	498	709	743	750	856
Isopropyl alcohol	2,069	1,674	1,938	1,521	1,936
Methyl alcohol	3,432	4,950	6,878	5,176	6,242
Perchloroethylene	533	705	734	679	669
Propylene oxide	814	1,194	1,756	1,524	1,823
Trichloroethylene	490	515	388	293	315
Urea	4,182	6,250	7,578	7,598	8,162
Vinyl acetate monomer	603	931	1,403	1,290	1,481
Vinyl chloride monomer[a]	6,395	11,895	14,786	12,173	13,719
Total	36,293	50,307	61,352	54,174	59,481

[a]Includes production of ethylene dichloride.

Source: *Synthetic Organic Chemicals.*

Table C–2.   Energy-Output Ratios by Chemical, Selected Years, 1967–76
(1,000 Btu per pound)

	1967	1971	1974	1975	1976
Acetic acid	12.8	11.3	10.2	9.9	9.5
Acetic anhydride	7.0	6.7	6.5	6.4	6.3
Acetone	3.4	3.0	2.6	2.5	2.4
Acrylonitrile	4.3	3.0	2.0	1.7	1.4
Adipic acid	14.5	13.4	12.6	12.3	12.1
Carbon disulfide	3.0	2.7	2.5	2.4	2.3
Ethyl alcohol	5.4	5.3	5.2	5.1	5.1
Ethyl chloride	2.2	2.1	2.1	2.1	2.1
Ethylene glycol	7.0	5.4	4.1	3.7	3.3
Ethylene oxide	1.4	0.6	0.1	-0.1	-0.3
Formaldehyde	2.0	1.1	0.4	0.15	-0.1
Hexamethylene diamine	34.0	29.0	25.2	23.9	22.7
Isopropyl alcohol	6.0	5.5	5.2	5.1	4.9
Methyl alcohol	7.0	6.8	6.7	6.7	6.7
Perchloroethylene	2.8	2.7	2.6	2.6	2.6
Propylene oxide	4.5	7.7	10.2	11.0	11.8
Trichloroethylene	2.0	3.5	4.7	5.1	5.5
Urea	4.0	3.4	2.9	2.8	2.6
Vinyl acetate monomer	4.3	5.1	5.8	6.0	6.2
Vinyl chloride monomer[a]	2.2	2.0	1.8	1.7	1.6

[a] Includes energy for ethylene dichloride.

Source: J. G. Myers, *Energy Consumption in Manufacturing*; see text, p. 137.

Index

About the Authors

John G. Myers is Associate Professor of Economics at Southern Illinois University at Carbondale, where he teaches courses on the economics of energy and the environment. Mr. Myers received the Ph.D. from Columbia University. He has specialized in energy and related concerns since 1972.

Leonard Nakamura is an economic consultant for The Conference Board and other organizations in New York City. He is an expert in industry studies and microeconomic theory. His work has spanned studies of energy, water pollution abatement, government regulation, productivity analysis, and input-output forecasting. His wide-ranging current research interests include business conditions analysis, social theory, environmental protection, and energy conservation.

About The Conference Board
and the Alliance to Save Energy

The Conference Board, established in 1916 as an independent nonprofit organization, performs research in the fields of economics, management, and policy development. Supported by over 4,000 associate organizations, including business and professional firms, government agencies, labor unions, and universities, the Board seeks, by providing objective information, to improve the quality and effectiveness of public and private sector leadership and to create broader understanding through conferences, seminars, workshops and published research reports.

The Alliance to Save Energy is a private, nonprofit, nonpartisan organization representing business, industrial, consumer, labor, and environmental concerns. The purpose of the Alliance is to raise public awareness in all economic sectors to the nature of the energy problem, and to stimulate action to improve the efficiency with which we use energy resources.